Aspergillus fumigatus

Contributions to Microbiology

Vol. 2

Series Editor *Axel Schmidt*, Wuppertal

KARGER

Basel · Freiburg · Paris · London · New York · New Delhi · Bangkok · Singapore · Tokyo · Sydney

Aspergillus fumigatus

Biology, Clinical Aspects and Molecular Approaches to Pathogenicity

Volume Editors *Axel A. Brakhage*, Darmstadt
Bernhard Jahn, Mainz
Axel Schmidt, Wuppertal

46 figures, 10 in color, and 21 tables, 1999

KARGER

Basel · Freiburg · Paris · London · New York ·
New Delhi · Bangkok · Singapore · Tokyo · Sydney

Contributions to Microbiology

formerly 'Concepts in Immunopathology' and
'Contributions to Microbiology and Immunology'

Axel Schmidt

PD Dr. med., FIBA
Bayer AG
Pharma Research Antiinfectives
Wuppertal;
Institute for Microbiology and Virology
University Witten-Herdecke, Germany

Library of Congress Cataloging-in-Publication Data
Aspergillus fumigatus: biology, clinical aspects, and molecular approaches to pathogenicity /
volume editors, Axel A. Brakhage, Bernhard Jahn, Axel Schmidt.
(Contributions to microbiology; vol. 2)
Includes bibliographical references and indexes. (hardcover : alk. paper)
1. *Aspergillus fumigatus.* 2. Aspergillosis. I. Brakhage, Axel A. II. Jahn, Bernhard.
III. Schmidt, Axel, 1962– . IV. Series.
[DNLM: 1. *Aspergillus fumigatus* – pathogenicity. 2. Aspergillosis – microbiology.
QW 180.5.D38 A8392 1999]
QR201.A85A86 1999
579.5′657–dc21
ISBN 3–8055–6714–6

Bibliographic Indices. This publication is listed in bibliographic services, including Current Contents® and Index Medicus.

Drug Dosage. The authors and the publisher have exerted every effort to ensure that drug selection and dosage set forth in this text are in accord with current recommendations and practice at the time of publication. However, in view of ongoing research, changes in government regulations, and the constant flow of information relating to drug therapy and drug reactions, the reader is urged to check the package insert for each drug for any change in indications and dosage and for added warnings and precautions. This is particularly important when the recommended agent is a new and/or infrequently employed drug.

Printed in Switzerland on acid-free paper by Reinhardt Druck, Basel
ISBN 3–8055–6714–6

In Remembrance of
Manfred W. Plempel (1930–1994)
Clinical Mycologist and Pioneer in Azole-Antifungal Research

Contents

Preface

In medicine, life-threatening systemic mycotic infections are gaining importance as opportunistic infections. This is due to the steadily increasing number of immunocompromised individuals: AIDS and cancer patients, organ transplant patients, or patients under intensive care or predisposing concomitant treatment regimens such as steroid therapy, therapy with other immunosuppressives, and antineoplastic and/or antibiotic chemotherapy. Further, fungal organisms can in the meantime also be attributed to nosocomial infections; pathogenic fungi belong to the fourth most isolated pathogens in intensive care units. Apart from systemic yeast infections, systemic mycoses can often be attributed to hyphomycetes, mostly to *Aspergillus* species, predominantly *A. fumigatus*. Most of these systemic *Aspergillus* infections are life-threatening with an extremely high mortality rate, even after appropriate antimycotic therapy regimens. This is mainly due to the often severe underlying immunosuppression of the host.

In general, infections/disorders caused by *Aspergillus* species, in particular *A. fumigatus*, can be divided into allergic reactions, generalized invasive and semi-invasive/restricted infections, and diseases resulting from mycotoxin exposure. All these aspects are discussed extensively and presented in this book which includes clinically oriented chapters, diagnostics, therapy, and new approaches in molecular biology.

This exciting project could not have been successful without the excellent editorial collaboration of my friends and colleagues Axel A. Brakhage (Darmstadt) and Bernhard Jahn (Mainz). I would like to thank both of them very much for constructive and effective discussion during this project and especially the cooperation with the inspiring ongoing research on *Aspergillus*-related topics as well as their amicable way of interaction and information exchange.

Witten-Herdecke, February 1999

Axel Schmidt
Series Editor

Editorial Preface

Fungal pathogens have begun to rival their bacterial counterparts as a significant problem in many settings. Death from fungal infection and bacterial infection are now almost equal in number in neutropenic patients. Besides the rise in occurrence of nosocomial yeast infections, infections with mold pathogens have emerged as an increasing risk faced by patients under sustained immunosuppression. Species of the *Aspergillus* family account for most of these infections and in particular *Aspergillus fumigatus* can be regarded as the primary mold pathogen. The improvement in transplant medicine and the therapy of hematological malignancies is often complicated by the threat of invasive aspergillosis. Specific diagnostics are still limited as are the possibilities of therapeutic intervention, leading to the disappointing fact that invasive aspergillosis is still associated with a high mortality rate that ranges from 30 to 90%.

However, the spectrum of diseases related to *A. fumigatus* is rather heterogeneous and exceeds that of a mere opportunistic pathogen: Colonization with restricted invasiveness is seen in the immunocompetent host, disseminated infections in the immunocompromised patient, and allergic disorders have long been known to be caused by *A. fumigatus* antigens.

While preparing this book, we realized that the heterogeneity of aspects which are important to be outlined in a book about *A. fumigatus* pathogenicity is probably comparable to the heterogeneity concerning the interests and backgrounds of our potential readers. Hence, our main objective to provide a concise and up-to-date report on *A. fumigatus* as the predominant human mold pathogen was guided by three main intentions.

First, we wished to provide the reader with a background on *A. fumigatus* biology, including physiology and taxonomy as a basis for further understanding of the more-detailed aspects related to the pathogenicity of *A. fumigatus*.

The second objective was to give a condensed overview of the medical aspects of *A. fumigatus*. Host defense mechanisms, clinical signs and underlying pathogenetic mechanisms of the different forms of *A. fumigatus*-related diseases are presented. The chapters on diagnostics and typing also include new molecular-based methods. The therapeutic strategies available against *A. fumigatus*-related diseases are summarised. In addition, information on the development of new antifungal drugs and susceptibility studies is given.

The question what makes *A. fumigatus* the predominant mold pathogen has not been resolved yet. Hence, our third intention was to give an overview about the current state of search for determinants of *A. fumigatus* virulence. Different chapters are dealing with the identification of potential virulence factors or physiological characteristics of *A. fumigatus* making it an aggressive opportunistic pathogen. Furthermore, the current knowledge on some potential targets for antifungal drugs like chitin synthases is presented. Most of the investigations which are summarised in these chapters use molecular genetics and protein chemistry as tools. In our view, these techniques represent the key technologies in the further elucidation of *A. fumigatus* pathogenicity. They are presented in various chapters, each covering a different aspect.

We are deeply indebted to the authors of the chapters in this volume whose intelligent and diligent efforts made this joint project possible. We are grateful to them for the care and energy with which they approached this work. We thank the series and volume editor Axel Schmidt for his enthusiasm when preparing this volume and for his excellent scientific cooperation and friendship. The editoral and production staff of S. Karger AG are gratefully acknowledged for their fruitful and professional collaboration.

Axel A. Brakhage/Bernhard Jahn
Darmstadt/Mainz

Brakhage AA, Jahn B, Schmidt A (eds): *Aspergillus fumigatus*.
Contrib Microbiol. Basel, Karger, 1999, vol 2, pp 1–4

J.B. Georg W. Fresenius and the Description of the Species *Aspergillus fumigatus*

A. Schmidt, D.I. Schmidt

Institut für Mikrobiologie und Virologie, Universität Witten-Herdecke, Deutschland

The species *Aspergillus fumigatus* was described in 1863 by Johann Baptist Georg Wolfgang Fresenius [1] as cited below

"*Aspergillus fumigatus* Fresenius

Mycelium ästig, sparsam septirt, ungefärbt. Sporentragende Fäden nach oben allmählich keulenförmig verdickt, in ein kugeliges, nach Entfernung der Sporen $^1/_{60}$ bis $^1/_{33}$ mm dickes Köpfchen ausgehend, unterhalb desselben bei durchfallendem Licht unter Wasser rauchgrau, nicht septirt, nur selten an der Basis oder auch in der Mitte mit einer Querwand. Verästigung der fructificirenden Hyphen, wie es scheint, nur selten; Figur 5, sowie das Figur 8 abgebildete Fragment könnte vielleicht hierher zu ziehen seyn. Eine so beträchtliche Ramification, wie in dem von Virchow mitgetheilten Fall von Bronchienverschimmelung (a. a. O. Fig. 3, d) ist mir bis jetzt noch nicht vorgekommen. Träger der Sporenketten länglich; der von ihnen bedeckte Theil des Köpfchens sepiabraun. Sporen $^1/_{400}$ mm gross, rund, einfach, glatt, grünlich. Die Fäden mit dem sporenlosen Köpfchen haben eine Länge von $^1/_6$–$^1/_3$ mm. Bei der Form von *Otis tarda* fand ich die längsten Hyphen auch $^1/_4$–$^1/_3$ mm messend.

Diese schöne, durch ihr Vorkommen interessante *Aspergillus*-Form hatte ich früher Gelegenheit, in mir zugekommenen Exemplaren aus der menschlichen Lunge, von Virchow und Pagenstecher herrührend, kennen zu lernen, und vor Kurzem bei einer Trappe (*Otis tarda*) des Frankfurter zoologischen Gartens zu beobachten, in deren Bronchien und andern Lufthöhlen Dr. Weinland sie aufgefunden hatte. Beide Fälle wurden von mir in den Sitzungen des mikroskopischen Vereins dahier besprochen; ich kann mich jedoch hier um so kürzer fassen, als über diese und eine nahe verwandte *Aspergillus*-Form Virchow im Archiv für pathol. Anat. Bd. IX, Heft 4, und Robin in seiner Hist. nat. des végétaux parasites bereits ausführlicher gehandelt haben. Die in Rede stehende Art ist dem *Aspergillus nigrescens* Robin nahe verwandt, auch in Beziehung auf den Standort ('sur des productions morbides tapissant les sacs aeriens d'un faisan'), unterscheidet sich aber durch grünliche Sporen (die Sporen von *A. nigrescens* werden *opaco-fusca* v. *nigra* genannt), und die Beschaffenheit der sporentragenden Hyphen; letztere werden bei der Robin'schen Art als aus mehren aneinandergereihten,

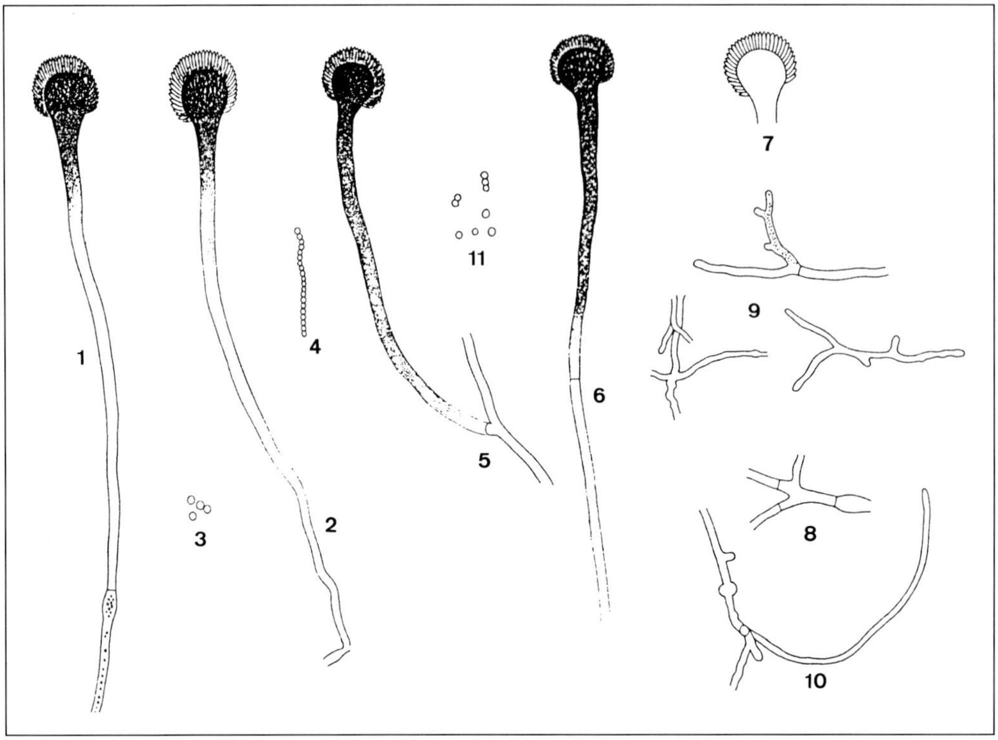

Fig. 1. "*Tafel X.* Figur 1–4 aus den Bronchien und Lufthöhlen der Trappe, 1–3 nach 350mal. Vergrösserung. 4 eine Sporenkette. Figur 5–11 aus der menschlichen Lunge; 9 u. 10 Theile vom Mycelium; sämmtliche Figuren 350mal vergrössert."

'Tafel X'. Figures 1–4 show isolates obtained from the bronchi and the lung of a bustard, figures 1–3: 350 × magnification. Figure 4: Single spore chain. Figures 5–11 show isolates obtained from human lung tissue; figure 9 and 10: parts from the mycelium. All figures × 350 magnification.

an den Berührungsstellen sogar etwas eingezogenen, articulirten Zellen bestehend beschrieben, während die fertilen Hyphen des *A. fumigatus* nicht septirt sind, vergl. die Abbildungen, besonders Figur 2, wo ein vollständiger Faden von seinem Ursprung aus dem Mycelium an dargestellt ist. So ist es hier die Regel; nur selten kommt eine Querwand vor, wie in Figur 6, oder in der besonders langen, $^1/_2$ mm messenden Hyphe Figur 1, in deren Mitte sich eine Auftreibung findet und wo sich unten in dem verschmälerten Theil eine zweite Querwand zeigt; aber eine Beschaffenheit der Fäden, wie bei *A. nigrescens* habe ich hier nie bemerkt. Von *A. glaucus* ist unsere Art so verschieden, dass sie nicht damit verglichen werden kann, und auch von den andern beschriebenen Arten stimmt keine genügend damit überein." (Fig. 1.)

'Mycelia branched, sparsely septated and colorless. Spore-bearing chains (=conidiophores) somewhat broader and club-shaped at the end, after removal of the spores ending in a thick head with a size of 0.015–0.03 mm (in diameter). Below the head (the conidiophore is) a smoky grey color if observed in a wet mount, not septated and only rarely having a septation at the base or in the middle. Hyphae which show fructification (=conidiophores) seem to be only rarely branched as shown in figure 5, and figure 8 where fragments of the fungus are shown. I could never observe an extensive branching as published by Virchow from an isolate obtained from a case of a (human) bronchial mould infection (fig. 3, d).

Spore-bearing elements (=conidiophores?) elongated; the part of the head bearing the spores is somewhat sepia-brown in color. Spores are 2.5 µm in size (diameter), globose, single celled, smooth, with a green color. The chains together with the sporeless head (=conidiophores?) have a length of 0.15–0.3 mm. In an isolate which I found in an infection of the bustard species *Otis tarda* I could observe the longest hyphae with a size of 0.25–0.3 mm (in length).

From my studies of isolates obtained from human lung infections by Virchow and Pagenstecher I had the chance to learn more about this fascinating species of *Aspergillus* which is also attractive because of its distribution. Recently, Weinland also isolated a strain from the bronchi and lung of a bustard which died at the zoological garden in Frankfurt. I discussed all these cases during meetings of the microscopy association (?). To make it short, I want to refer to this species and a closely related species of *Aspergillus* as already described by Virchow in 'Arch Pathol Anat', vol IX, No. 4 and by Robin in 'Hist Nat Végétaux Parasites'. The species I describe is closely related to *Aspergillus nigrescens* Robin also as far as its distribution is concerned ('sur des productions morbides tapissant les sacs aeriens d'un faisan'). It differs from this species because of its green pigmentation of the spores [the spores of *A. nigrescens* are called (*A.* ?) *opaco-fusca* v. *nigra* (?)]. Further, the species as described by Robin shows septated spore-bearing hyphae (=conidiophores) with constrictions at the septation site. In contrast, *A. fumigatus* has no septated fertile hyphae (=conidiophores) as is shown especially in figure 2 which shows a whole hyphae beginning at the (basal) mycelium. This occurs regularly in *A. fumigatus;* only rarely can septation be observed as shown in figure 6. Septation can also be seen in the extremely long hyphae with a length of 1 mm as shown in figure 1. In this case the hyphae was thickened in the middle, and in the thinner part towards its base a second septum could be observed. I could never observe (frequent) septations (evident) in *A. nigrescens.* The species we describe here is very different to *A. glaucus* and also from all other (*Aspergillus*) species which have been described so far.'

(The translation was performed as strictly as possible in accordance with the German text.)

J.B. Georg W. Fresenius [2] was born in Frankfurt a.M., Germany, on 25.9.1808 and also died there on 1.12.1866 of pneumonia. He studied medicine in Heidelberg, Würzburg and Giessen (all Germany) and finished his doctorate thesis (MD) in Giessen in 1829. Fresenius started his career as a physician and surgeon in Frankfurt a.M. in the same year. He became a university lecturer for botany at the 'Senckenbergischen medicinischen Institut' (this institute specialized in botany – especially research on medicinal herbs – though astronomical research was also performed) in Frankfurt a.M. in 1831 and became the director of the botanical gardens of Frankfurt a.M. in the same year. He was awarded a professorship in botany in 1863.

The monograph 'Beiträge zur Mykologie' (Contributions to Mycology) was published by Fresenius as a dedication for the centennial celebrations of the Senckenberg foundation ('Senckenbergische Stiftung') on 18.8.1863 in Frankfurt a.M. It has 132 pages and further contains 13 excellent lithographic figures (camera lucida) and is divided into three parts which were published together as a monograph. The three parts are identical concerning their structure and are a list of descriptions of fungal species. Part 1 (pp. 1–38) was already published in 1850 and part 2 (pp. 39–80) in 1852 on its own. The whole monograph, including the third part (pp. 81–132; this part also contains the description of the species *A. fumigatus* on pp. 81–82 and lithographic 'Tafel X') was published in 1863 and contains an overall description of 132, mostly new or revised fungal species and/or genera.

References

1 Fresenius G: Beiträge zur Mykologie. Frankfurt a.M., Brönner, 1863, pp 81–82.

2 Kallmorgen W: Siebenhundert Jahre Heilkunde in Frankfurt am Main. Frankfurt a.M., Diesterweg Verlag, 1936, pp 268–269.

Priv.-Doz. Dr. med. Axel Schmidt, FIBA, Facharzt für Mikrobiologie und Infektionsepidemiologie, Bahnstrasse 118, D–42327 Wuppertal (Germany)
Tel./Fax +49 202 780961, E-Mail axel780961@aol.com

Brakhage AA, Jahn B, Schmidt A (eds): *Aspergillus fumigatus*.
Contrib Microbiol. Basel, Karger, 1999, vol 2, pp 5–20

The Genus *Aspergillus* with Special Regard to the *Aspergillus fumigatus* Group

Robert A. Samson

Centraalbureau voor Schimmelcultures, CBS, Baarn, The Netherlands

Raper and Fennell [1] accepted 132 *Aspergillus* species subdivided into 18 groups when they monographed the genus *Aspergillus*. For that time the genus and species concepts were well circumscribed and this monograph is still valuable and used. However, many new species have been described and the relationship of *Aspergillus* with the teleomorph states has been elucidated. In addition, several molecular studies have shed a light on the phylogeny of the genus *Aspergillus*. Samson [2–4] provided a compilation of the species and varieties described since 1965 with a critical review of the validity of the published taxa. The genus now contains more than 180 species, with about 70 named teleomorphs.

The current classification of *Aspergillus* and its teleomorphs is greatly influenced by the outcome of three international workshops on *Penicillium* and *Aspergillus* systematics [5–7]. Important contributions to the taxonomy of the genus came from multidisciplinary approaches including biochemistry, molecular biology and serology. Furthermore, strict adherence to the rules of the 'International Code of Botanical Nomenclature' (ICBN) and some new nomenclatoral approaches have improved the naming of Aspergilli. This is in contrast with many other fungal genera where much confusion exists about the correct names.

Today species of *Aspergillus* are important organisms in industry and many fields of applied research including medical mycology. This paper reviews the current taxonomic concepts and classification of *Aspergillus* and their teleomorphs with emphasis on the section of *Aspergillus fumigatus* and related taxa.

Classification of *Aspergillus* and Its Teleomorphs

Aspergillus is a form genus consisting of species whose teleomorphs have been described in eight different genera. Therefore, *Aspergillus* could have split

Table 1. Infrageneric taxa of the genus *Aspergillus*

Subgenus	Section
Aspergillus	*Aspergillus* (*A. glaucus* group), *Restricti* (*A. restrictus* group)
Fumigati	*Fumigati* (*A. fumigatus* group), *Cervini* (*A. cervinus* group)
Ornati	(*A. ornatus* group)
Clavati	*Clavati* (*A. clavatus* group)
Nidulantes	*Nidulantes* (*A. nidulans* group), *Versicolor* (*A. versicolor* group), *Usti* (*A. ustus* group), *Terrei* (*A. terreus* group), *Flavipedes* (*A. flavipes* group)
Circumdati	*Wentii* (*A. wentii* group), *Flavi* (*A. flavus* group), *Nigri* (*A. niger* group), *Circumdati* (*A. ochraceus* group), *Candidi* (*A. candidus* group), *Cremei* (*A. cremeus* group), *Sparsi* (*A. sparsus* group)

Group names following the classification of Raper and Fennel [1] are given in parentheses.

up in several entities, but the *Aspergillus* anamorphs are mostly considered as a taxonomic unit and therefore have been viewed as naturally related forms [1, 8–11]. Additional data obtained with different research techniques have appeared since the last comprehensive monograph [1] including morphological [2, 11], biochemical [12, 13], ecological, genetic [14–17], and molecular genetic characteristics [18–22].

Several sections of the genus, which remained problematic for identification, have been reinvestigated, including subgenus *Aspergillus* section *Aspergillus* (= the *A. glaucus* group = *Eurotium*) [11, 23, 24], section *Nigri* [11, 25, 26], section *Circumdati* [27], section *Nidulantes* = *Emericella* [28, 29], section *Versicolores* [30], section *Flavi* [31, 32], section *Fumigati* [12, 21, 33, 34], sections *Cremei* and *Wentii* [35], and section *Restricti* [36].

The genus is now subdivided into six subgenera and 18 sections (table 1) [37] and these sections largely correspond to 18 groups proposed by Raper and Fennell [1]. This infrageneric classification into 'groups' as proposed by Raper and Fennell [1] had no nomenclatural status under the ICBN.

Aspergillus has teleomorphs placed in nine genera (table 2). The most common teleomorph genera are *Eurotium, Neosartorya* and *Emericella*. Particularly in the last genus many new species have been described [4, 38]. A list of recently described species of *Neosartorya* is given under section *Fumigati*. Some genera such as *Hemicarpenteles* and *Warcupiella* have been found to be distantly related to other *Aspergillus* teleomorphs. A more detailed study of sufficient numbers of isolates of representatives of these genera should warrant their phylogenetic position.

The majority of *Aspergillus* teleomorphs are homothallic. Heterothallic taxa can be found in three species of *Neosartorya: N. fennelliae* Kwon-Chung

Table 2. Ascomycetous genera with an *Aspergillus* anamorph

Section *Aspergillus:*	*Eurotium* Link:Fr., *Dichlaena* Mont. and Durieu, *Edyuillia* Subram.
Section *Fumigati:*	*Neosartorya* Malloch and Cain
Subgenus *Ornati:*	*Warcupiella* Subram, *Sclerocleista* Subram, *Hemicarpenteles* Sarbhoy and Elphick
Section *Nidulantes:*	*Emericella* Berk. and Br.
Section *Flavipedes:*	*Fennellia* Wiley and Simmons
Section *Circumdati:*	*Petromyces* Malloch and Cain
Section *Cremai:*	*Chaetosartorya* Subram.

Edyuillia = *Eurotium; Hemicarpenteles* is based on *H. paradoxus,* which is distantly related to *Aspergillus* teleomorphs.

and Kim, *N. spathulata* Takada and Udagawa, *N. udagawae* Horie et al., and *Emericella heterothallica* (Kwon-Chung et al.) Malloch and Cain. Further, some *Emericella* species are heterothallic. In some *Emericella* species such as *E. desertorum* Samson and Mouchacca and *E. pleuriseminata* Stchigel and Guarro, the teleomorphs are not known to produce an *Aspergillus* anamorph, but these fungi possess sufficient characters to justify their placement in these genera [3, 4]. Together with the morphological characters of the anamorph, the teleomorph species are easy to identify. Particularly, the ornamentation of the ascospores, which can be readily seen in both light microscopy and scanning electron microscopy (SEM) is important to identify the taxa. It is therefore not surprising that, for example, in the genera *Neosartorya* and *Emericella* several new species have been described primarily based on unique ascospore ornamentation. For the examination of the ornamentation of thick-walled ascospores in the unfixed, air-dried state, the use of SEM can be useful but this must be applied with caution when examining conidia and other structures, which are less resistant to shrinkage. In addition, the molecular results obtained by Geiser et al. [34] also implicate that the *Aspergillus* taxonomy as proposed by Kozakiewicz [11] using primarily SEM and spore ornamentation are not reflected in the molecular phylogeny. Differences in pigmentation and spore ornamentation may have only a simple genetic basis and to be susceptible for convergent evolution.

Nomenclature

Nomenclatoral problems with the genus *Aspergillus* have been solved. These problems were caused by neglecting the priority of older names and the failure to specify dried types for accepted or newly described species according to the rules of the ICBN [39]. Samson and Gams [40] and Kozakiewicz [11] typified the majority of the *Aspergillus* names used by Raper and Fennell [1].

Many *Aspergillus* species have an ascomycete teleomorph (table 2), but the names used were not in accordance with article 59 of the ICBN because Raper and Fennell [1] consistently retained the *Aspergillus* names for both the teleomorph and anamorph. Samson and Gams [40] and Kozakiewicz [11] carried out the necessary nomenclatoral changes [36, 41].

Stabilization of Aspergillus *Nomenclature*

Hundreds of *Aspergillus* names have been described since Micheli erected the genus in 1729. Many of these names are based on a single dried specimen which has been collected in nature and of which no culture has been made or deposited. These specimens with obscure names are deposited in many herbaria and the names could easily be picked up again. To avoid confusion of *Aspergillus* nomenclature two procedures were recently introduced.

In the case of *A. niger*, a name of major economic importance, the conservation of the name was formally proposed by Kozakiewicz et al. [42]. Also, the well-known name *A. nidulans*, a species used in a wide range of significant genetic studies, was proposed for conservation because the name *A. nidulans* was threatened by the legitimate name *A. nidulellus* [40]. This anamorph name had to be established for the teleomorph *Emericella nidulans*, following a change to the rules adopted in 1981.

A second approach for stabilizing *Aspergillus* nomenclature is the protection of names in current use. Pitt and Samson [43], on behalf of the International Commission on *Penicillium* and *Aspergillus* (ICPA, a commission of the 'International Union of Microbiological Societies') proposed a list of species names in current use in the family *Trichocomaceae* (Fungi, *Eurotiales*) including *Aspergillus* and its teleomorphs. In order to protect *Aspergillus* names in current use from being threatened or displaced by names that are no longer in use, and in order to eliminate uncertainties regarding their application spelling, gender, typification date and place or valid publication, a published lists of names including names of 186 Aspergilli and 72 teleomorphs with an *Aspergillus* anamorph in current use were proposed to the 'XV International Botanical Congress in Tokyo' for protection. Although this list was not formally accepted at the Congress, the list of 'Names in Current Use', was given special status [39]. This was a significant step towards the goal of providing stable nomenclature for this important genus and *Penicillium* into the foreseeable future. Samson and Pitt [7] carried the process one step further, by providing a much needed list of synonyms of those species. Taken from several sources, the list aims to be definitive where possible, and explanatory where it is not. Although this list will not have any formal nomenclatoral satus, it should be of great assistance to all those who work with and use *Aspergillus* taxonomy.

Morphological and Cultural Criteria

The *Aspergillus* conidiophore is easy to recognize and this allows a rapid and easy identification when sporulating structures have been formed on the natural substrate or in clinical specimens. The conidiophore with a typical aseptate stipe terminating in a vesicle, on which the conidiogenous cells – phialides singly or on metulae – are borne, is still a primary morphological criterion for delimitation of the genus. Only in cases where the isolate is atypical, cross walls in the stipe might be found and the metulae and phialides show branched or swollen structures. The color of the conidial heads is also an important identification character, together with characteristics such as Hülle cells and sclerotia. This color which is still used for the group classification and is indicative for the section and group is dependent on the degree of sporulation. The color might also be deviant because of the medium. Even on standardized Czapek agar pigmentation can be atypical. It is therefore important to strictly follow the preparation of the media. The addition, supplemental solution containing copper and zinc can solve this atypical growth [44]. In clinical isolates, e.g. in the case of *A. fumigatus*, or in strains frequently used in biotechnological applications, the typical conidial pigmentation might be absent due to the formation of abundant aerial mycelium. Many species produce sclerotia on agar media when freshly isolated but unfortunately the production of these vegetative bodies can easily be lost and therefore this albeit significant morphological character is not often used.

Keys for Identification and Reference Cultures

Practical laboratory guides for identification of common Aspergilli which are primarily based on morphology and colony characters have been published by Klich and Pitt [10], Tzean et al. [45], Samson et al. [44], and Pitt and Hocking [46]. Klich and Samson [47] give detailed descriptions of a set of reference cultures. This set of reference cultures was developed by ICPA to improve the accuracy of *Aspergillus* identification. The isolates were selected to represent a wide variety of morphological and physiological characteristics of the genus. Using these living cultures, individuals can observe these characteristics under their own lab conditions. Species included in the set were

A. candidus, A. clavatus, A. flavus, A. fumigatus, A. niger, A. ochraceus, A. penicillioides, A. terreus, A. versicolor and two species with *Aspergillus* anamorphs, *Emericella nidulans* and *Eurotium amstelodami*. This culture set which is available from the major culture collections provides basic information on the genus in order to facilitate research and identification. The descriptions of the cultures are available on the Internet (http://www/cbs.knaw.nl/icpa).

Media for Detection and Identification

Most Aspergilli can readily be identified when cultivated on Czapek and malt extract agars. Samson and Pitt [5] recommended Czapek yeast agar (CYA) and malt extract agar (MEA) for most conidial species. Also Schmidt and Wolff [48] found that isolates of *Aspergillus* should be best identified using a standardized Czapek agar. For teleomorphic species, cornmeal agar, oatmeal agar and CBS malt agar are recommended. For xerophilic species, it is important to use low water activity media such as Czapek agar or CYA with 20% sucrose or malt yeast extract agar with 40% sucrose. Isolates of *Eurotium* or section *Restricti* are often misidentified because wrong media were used. Several isolates reported in the literature as *A. fumigatus* are *A. penicillioides* or *A. restrictus. Aspergillus* cultures should be incubated for 5–7 days at 25 °C for anamorphic species and 10–14 days for most teleomorphic species. For the detection and identification of *A. fumigatus* 37 °C can be used because the isolates will grow faster at this temperature than at 25 °C.

Sabouraud agar, which is commonly used in medical microbiology laboratories, is not considered to be suitable for isolation and identification. Although *A. fumigatus* isolates grow well at this medium, many other Aspergilli will not or will develop only atypically on this medium.

Some specialized media are used for the rapid recognition of potentially aflatoxigenic Aspergilli. For the detection of *A. flavus* and *A. parasiticus* within 3 days 'Aspergillus Differential Medium' [49] or 'Aspergillus *Flavus* and *Parasiticus* Agar' [50] can be used. Another medium containing the antibiotic bleomycin helps to distinguish *A. parasiticus* from *A. sojae* [51]. On this medium, growth of both species is somewhat reduced, but *A. sojae* produces very restricted colonies while *A. parasiticus* colonies are at least 3 mm in diameter within 6 days. These selective media are especially appropriate for the rapid detection of mycotoxinogenic isolates in food commodities or indoor environments.

For the detection of Aspergilli from air or indoor environments, malt extract agar or dichloran 18% glycerol agar (DG18) incubated at 25 °C are recommended. However, for the rapid detection of airborne *A. fumigatus*, incubation at 37 °C can be very helpful. Media containing Rose Bengal should be treated with caution because this growth inhibitor can be toxic to moulds when the media are exposed to light.

New Approaches for Taxonomic Purposes

Several new approaches have been tested to solve taxonomic problems in *Aspergillus* or to elucidate phylogeny. These new techniques can be divided into biochemical and molecular studies. Studies on secondary metabolites, ubiquinones and many molecular studies for taxonomic or phylogenetic studies are reviewed by Samson [4]. Ubiquinone systems (Q-9 and Q-10), often in combination with electrophoretic comparison of enzymes, have been used by several Japanese researchers [13, 52] and this approach indicates relationships between groups, but cannot be used for identification purposes. It, however, may also help to clarify the degree of relatedness among important species. Clinical and nonclinical isolates of *A. fumigatus* showed homogeneity, while glucose-6-phosphate dehydrogenase and glutamate dehydrogenase were key enzymes for differentiating isolates of the species [52].

For the detection of Aspergilli various immunoassay systems and production of monoclonal antibodies, which identify unique epitopes, have been developed. Practical serological methods for the detection of heat-stable extracellular polysaccharides, an ELISA method and a latex agglutination method were developed [53]. For example, an antigen latex agglutination test for the detection of Aspergilli in the serum and/or liquor cerebrospinalis of patients with invasive aspergillosis was modified for the detection of foodborne Aspergilli and Penicillia [54]. Until now, no immunoassays are reported which are specific for species or strains and cannot be used for intraspecies identification.

Profiles of secondary metabolites, including mycotoxins, have proven to be important for the taxonomy of *Penicillium* and are also of great use in *Aspergillus* systematics. For example, Samson et al. [12] distinguished nine taxa within the genus *Neosartorya* by ascospore morphology and the profile of secondary metabolites. Frisvad and Samson [33] carried out a biochemical and morphological study of 27 isolates of *A. fumigatus* and related taxa. The secondary metabolite production by *A. fumigatus* and described varieties was homogeneous: tryptoquivalins, fumigaclavine A, verruculogen, other fumitremorgins, and fumagillin were produced by all isolates tested, fumigatin- and sulochrin-like metabolites were produced less frequently while gliotoxin and helvolic acid were not detected (table 3). The examined isolates of *A. brevipes, A. viridinutans, A. duricaulis* and *A. unilateralis* also produced unique profiles

Table 3. Species of *Neosartorya* with an *Aspergillus* anamorph

Taxa as accepted by Raper and Fennell [1] but combined to the ascomycete genus by other authors
- *Neosartorya aurata* (Warcup) Malloch and Cain
- *Neosartorya aureola* (Fennell and Raper) Malloch and Cain
- *Neosartorya fischeri* Wehmer
- *Neosartorya glabra* (Fennell and Raper) Kozakiewicz
- *Neosartorya quadricincta* (Yuill) Malloch and Cain
- *Neosartorya spinosa* (Raper and Fennell) Kozakiewicz
- *Neosartorya stramenia* (Novak and Raper) Malloch and Cain

Species described after Raper and Fennell [1]
- *Neosartorya botucatensis* Horie et al. (from soil, Brazil)
- *Neosartorya delicata* Kong (from fruit, China)
- *Neosartorya fennelliae* Kwon-Chung and Kim* (from eyeballs of a rabbit)
- *Neosartorya hiratsukae* Udagawa et al. (from pasteurized aloe beverage, Japan)
- *Neosartorya multiplicata* Yakuchi et al. (from soil, Taiwan)
- *Neosartorya paulistensis* Horie et al. (from soil, Brazil)
- *Neosartorya primulina* Udagawa and Toyazaki (from canned oolong tea, Japan)
- *Neosartorya pseudofischeri* Petersen (from cardboard and many other substrates [21])
- *Neosartorya spathulata* Takada and Udagawa* (from soil, Formosa)
- *Neosartorya tatenoi* Horie et al. (from soil, Brazil)
- *Neosartorya udagawae* Horie et al.* (from soil, Brazil)

* Heterothallic species, all other species in this list are homothallic.

of secondary metabolites different from those in *A. fumigatus.* The taxa of sect. *Fumigati* as proposed by Raper and Fennell [1] are therefore well defined.

Several recent molecular studies on section of *Aspergillus* were carried out and these are described in Samson and Pitt [7]. One important study is by Peterson [55] who sequenced the D1 and D2 region of the large subunit ribosomal RNA (lsu-DNA) genes from more than 200 taxa of *Aspergillus.* These molecular data showed that the maximum parsimony tress were not compatible with the current morphological taxonomy of the subgeneric delimitation. The three major lineages were the subgenera *Aspergillus, Nidulantes* and *Fumigati*. The last subgenus also includes species normally placed in section *Clavati*.

Section *Fumigati*

Species of the section are characterized by smooth-walled stipes, which are often greenish. Conidiophores are uniseriate with subclavate vesicles producing mostly compactly columnar heads, which are commonly gray-green to dark-blue-green. Phialides are flask-shaped and confined to the apical part. Hülle cells and sclerotia are unknown. A teleomorph *Neosartorya* Malloch and Cain [57] can be present in a homo- or heterothallic form. Cleistothecia in white or yellow colors occur, globose with a distinct mycelial wall. The globose to subglobose asci contain 8 ascospores. Ascospores are one-celled, globose to ellipsoidal/lenticular and covered with various ornamentation patterns.

The type species *A. fumigatus* was described by Fresenius [56].

Anamorphic species are: *A. brevipes* Smith, *A. duricaulis* Raper and Fennell, *A. fumigatus* Fresenius, *A. unilateralis* Thrower, and *A. viridinutans* Ducker and Thrower.

The taxonomy of the members of this section is important because they play a significant role as human pathogens and as food contaminants. Although other members are important in medical mycology, especially two species from section *Fumigati, A. fumigatus* and *N. pseudofischeri*, are known to cause human disease [21, 58–61].

A. fumigatus is morphologically more variable than described by Raper and Fennell [1]. Conidia may be rough-walled to nearly smooth, globose to ellipsoidal. Color mutants occasionally occur. Phialides may be septate – described as *A. phialiseptus* – and conidial heads may be somewhat radiate as in *A. fumigatus* var. *acolumnaris*. Varsney and Sarbhoy [62] even raised the variety *acolumnaris* to species level on the basis of subglobose heads instead of columnar heads. Clinical isolates, in particular, can be markedly abnormal [63, 64] and are often more floccose with less conidia (fig. 1). Such isolates cannot be separated from the typical saprophytic isolates from substrates such as soil or foodstuffs [52, 65–67]. Schmidt and Wolff [48] carried out a detailed morphological study of 47 clinical isolates and found that 43 isolates had a high morphological similarity. Only four strains were atypical differing in slower growth, conidia pigmentation, deviating phialide and vesicle shape and conidial size and shape. All 47 isolates showed faster growth at 37 and 42 °C than at 25 °C.

The ex-type isolate of *A. fumigatus* var. *ellipticus* differs from typical *A. fumigatus* isolates by its smooth-walled, ellipsoidal conidia and therefore Kozakiewicz [11] recognized it at specific rank. The isolate is however biochemically identical with *A. fumigatus* var. *fumigatus* [33, 65]. In addition, Peterson [21] made nuclear DNA complementary measurements and found 93% NDNA relatedness between the type isolates of both varieties and concluded that the name *A. neoellipticus* proposed by Kozakiewicz [11] should be treated as a synonym of *A. fumigatus*.

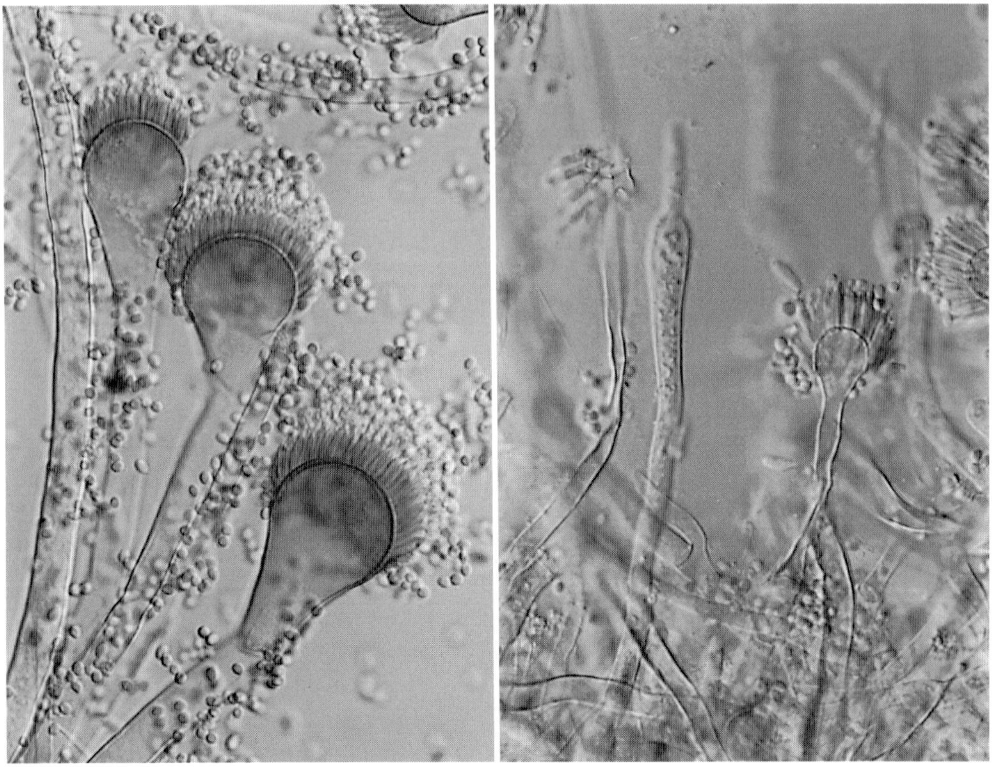

a b

Fig. 1. Conidiophores of *Aspergillus fumigatus*. *a* A typical conidiophore with uniseriate phialides. *b* Conidiophore with atypical phialides (magnification × 570).

Geiser et al. [34] examined the evolutionary relationships among the members of section *Fumigati* inferred from partial β-tubulin and hydrophobin DNA sequences. The data sets in combination showed that the morphological and secondary metabolites characteristics used in current taxonomic schemes were not strongly correlated with phylogeny. Taxa inferred to be closely related based on sequence data share certain morphological characters but these characters such as cleistothecium color and spore ornamentation can also be found in other clades. Their results implicate that the *Aspergillus* taxonomy proposed by Kozakiewicz [11] using SEM and the spore ornamentation are not reflected in the molecular phylogeny. Geiser et al. [34] further suggested that differences in pigmentation and spore ornamentation may have only a simple genetic basis and to be susceptible for convergent evolution.

Frisvad and Samson [unpubl. data] investigated the species of section *Clavati* and found that they were closely related on the basis of their morphology, physiology and production of secondary metabolites. Clustering, principal coordinate analysis and correspondence analysis showed that the taxa in *Clavati* could be placed in four distinct species. Some of the secondary metabolites found have also been detected in species of the genus *Neosartorya* and the section *Fumigati*, which indicate that section *Fumigati* and *Clavati* are closely related. Peterson [55] has confirmed the relationship between the two sections by sequence data.

A few new species in section *Fumigati* have been recently described but their taxonomic identity should be questioned. One species, *A. arvii* Aho et al. [68] isolated from liver lesions of a dairy cow in Finland, was considered different because of its stable brown colony color and less profound ornamentation of the conidia. However, both characteristics usually fall within the range of variation of *A. fumigatus.* Aho et al. [68] also considered their new species as identical with *A. fumigatus* var. *fulvi-ruber* Rai et al. [69], but there is no indication in their publication that they examined the type culture of this variety. Without a proper type indication *A. arvii* is also invalidly published.

Horie et al. [70] described one isolate from Venezuelan soil, *A. fumisynnematus* as a new species because this isolate produces synnemata with smaller conidial heads and short conidiophores. It further has less ornamented conidia when compared with *A. fumigatus.* The taxonomic identity of this species remains doubtful because no material is available for study.

Neosartorya Teleomorphs

Neosartorya species and in particular *N. glabra, N. pseudofischeri* and *N. spinosa* are important taxa, because they occur as heat-resistant contaminants in food and beverages which are mostly associated with fruit [44, 46]. Several new species are described, some from beverages, but mostly they occur in soil.

Geiser et al. [34] found that *A. fumigatus* and *N. fischeri* show a very close relationship, indicated by their molecular data but also by the production of the same secondary metabolites as reported by Frisvad and Samson [33]. Therefore, they suggested that *N. fischeri* has potential as a sexual model for studying *A. fumigatus.*

Raper and Fennell [1] accepted five species and three varieties of *Neosartorya* as the *A. fischeri* series on the basis of ascospore ornamentation and color of the ascomata. Malloch and Cain [57] listed 10 taxa, including three other teleomorphs in *Aspergillus*, later reclassified by Samson [2] in the teleomorphic genera *Hemicarpenteles* and *Chaetosartorya*. In addition to the taxa accepted by Raper and Fennell [1], three heterothallic species, *Neosartorya fennelliae, N. spathulata* and *N. udagawe*, and eight homothallic taxa have been described (table 3).

Horie [71] gave an overview of the taxonomy while Yaguchi et al. [72] provided a key to 14 species. In their study of *A. fumigatus* and related teleomorphs, Geiser et al. [34] found that a very close relationship was inferred from yellow-colored *N. aurata* and *N. stramenia*, despite the difference in ascospore ornamentation.

Peterson [21] examined the type material of *Aspergillus fischeri* var. *thermomutatus* and raised the anamorph to species as *A. thermomutatus* and described the teleomorph *N. pseudofischeri* as new species. Ascospores of this new species are distinguished by nonanastomosing ridges with triangular flaps. Several strains of *N. pseudofischeri* are from clinical origin but it is apparently not common as a human pathogen and certainly does not appear as a particularly virulent opportunistic fungal pathogen. Udagawa et al. [73] consider *N. pseudofischeri* a common potential fungal pathogen in indoor environments because they found this species when they analyzed house dust samples from apartments located around Kobe City, Japan. This teleomorph species could be isolated using a selective isolation method, using sieved dust samples immersed in 65% alcohol for 8 min, washed and plated out on potato dextrose agar.

Peterson [21] agreed with Kozakiewicz [11], who gave species status to *N. spinosa* and *N. glabra*, taxa that had previously been varieties of *N. fischeri*. These authors also suggested that *N. spinosa* might contain strains of more than a single species. The rDNA sequences of multiple strains of *N. spinosa* fall into two categories: those strains whose sequences are identical to that of the ex-type isolate, and another group with sequences closely related to *N. spathulata*. There are two genetically distinct groups within the current morphological concept of *N. spinosa*, and they appear to warrant species status.

Conclusion

Although the taxonomic concept of most *Aspergillus* species is clear and reliable identification is feasible, recent molecular studies show that the infrageneric classification should be reshuffled.

Some revisions of particular species or groups of species are carried out but have not provided the clear taxonomic classification to allow a simple and practical identification procedure. This is especially so in sections such as *Nigri* and *Flavi*, which contain species, which are important for biotechnological applications. Furthermore, in other sections, such as *Circumdati*

(= *A. ochraceus* group in accordance to Raper and Fennell [1]) with significant toxinogenic species, a revision is also urgently required. For the identification of *A. fumigatus* and its related species the procedures are simple and allow rapid recognition of the isolates.

References

1 Raper KB, Fennell DI: The Genus *Aspergillus*. Baltimore, Williams & Wilkins, 1965, pp 1–686.

2 Samson RA: A compilation of the *Aspergilli* described since 1965. Stud Mycol 1979;18:1–38.

3 Samson RA: Current taxonomic schemes of the genus *Aspergillus* and its teleomorphs; in Bennett JW, Klich MA (eds): Aspergillus: The Biology and Industrial Applications. London, Butterworth, 1992, pp 353–388.

4 Samson RA: Current systematics of the genus *Aspergillus*; in Powell KA, Renwick A, Peberdy JF (eds): The Genus *Aspergillus*: From Taxonomy and Genetics to Industrial Application. London, Plenum Press, 1994, pp 261–276.

5 Samson, RA, Pitt JI: Advances in *Penicillium* and *Aspergillus* Systematics. New York, Plenum Press, 1986.

6 Samson RA, Pitt JI: Modern Concepts in *Penicillium* and *Aspergillus* Classification. New York, Plenum Press, 1990.

7 Samson RA, Pitt JI: Integration of modern taxonomic methods for *Penicillium* and *Aspergillus* classification. Dordrecht, Kluwer, 1999.

8 Thom C, Church MB: The Aspergilli. Baltimore, Williams & Wilkins, 1926, pp 1–272.

9 Thom C, Raper KB: A Manual of the Aspergilli. Baltimore, Williams & Wilkins, 1945.

10 Klich MA, Pitt JI: A Laboratory Guide to Common *Aspergillus* Species and Their Teleomorphs. North Ryde, CSIRO, Division of Food Processing, 1988, pp 1–116.

11 Kozakiewicz A: *Aspergillus* species on stored products. Mycol Papers, 1989;161:1–188.

12 Samson RA, Nielsen PV, Frisvad JC: The genus *Neosartorya*, differentiation by scanning electron microscopy and mycotoxin profiles; in Samson RA, Pitt JI (eds): Modern Concepts in *Penicillium* and *Aspergillus* Classification. New York, Plenum Press, 1990, pp 455–467.

13 Kuraishi H, Itoh M, Tsuzaki N, Katayama Y, Yokoyama T, Sugiyama J: The ubiquinone system as a taxonomic aid in *Aspergillus* and its teleomorphs; in Samson RA, Pitt JI (eds): Modern Concepts in *Penicillium* and *Aspergillus* Classification. New York, Plenum Press, 1990, pp 395–406.

14 Yamatoya K, Sugiyama J, Kuraisi H: Electrophoretic comparison of enzymes as a chemotaxonomic aid among *Aspergillus* taxa: (2) *Aspergillus* sect. *Flavi*; in Samson RA, Pitt JI (eds): Modern Concepts in *Penicillium* and *Aspergillus* Classification. New York, Plenum Press, 1990, pp 395–405.

15 Cruickshank RH, Pitt JI: Isoenzyme patterns in *Aspergillus flavus* and closely related species; in Samson RA, Pitt JI (eds): Advances in *Penicillium* and *Aspergillus* Systematics. New York, Plenum Press, 1990, pp 259–265.

16 Egel DS, Cotty PJ, Elias KS: Relationships among isolates of *Aspergillus* sect. *Flavi* that vary in aflatoxin production. Phytopathology 1994;84:906–912.

17 Cotty PJ, Bayman P, Egel DS, Elias KS: Agriculture, aflatoxins and *Aspergillus*; in Powell KA, Renwick A, Perberdy JF (eds): The Genus *Aspergillus*, from Taxonomy and Genetics to Industrial Applications. New York, Plenum Press, 1994, pp 1–28.

18 Dupont J, Dutertre M, Lafay J-F, Roquebert M-F, Brygoo Y: A molecular assessment of the position of *Stilbothamnium* in the genus *Aspergillus*; in Samson RA, Pitt JI (eds): Modern Concepts in *Penicillium* and *Aspergillus* Classification. New York, Plenum Press, 1990, pp 335–342.

19 Chang J-M, Oyaizu H, Sugiyama J: Phylogentic relationships among eleven selected species of *Aspergillus* and associated teleomorphic genera estimated from 18S ribosomal RNA partial sequences. J Gen Appl Microbiol 1991;37:289–308.

20 Berbee ML, Taylor JW: Two ascomycete classes based on fruiting body characters and ribosomal DNA sequences. Mol Biol Evol 1992;9:278–284.

21 Peterson SW: *Neosartorya pseudofischeri* sp. nov. and its relationship to other species in *Aspergillus* section *Fumigati*. Mycol Res 1992;91:547–554.

22 Berbee ML, Yoshimura J, Sugiyama J, Taylor JW: Is *Penicillium* monophyletic? An evaluation of phylogeny in the family *Trichocomaceae* from 18S, 5.8S and ITS ribosomal DNA sequence data. Mycologia 1995;87:210–222.

23 Blaser P: Taxonomische und physiologische Untersuchungen über die Gattung *Eurotium* Link ex Fries. Sydowia 1975;28:1–49.

24 Pitt JI: Nomenclatoral and taxonomic problems in the genus *Eurotium*; in Samson RA, Pitt JI (eds): Advances in *Penicillium* and *Aspergillus* Systematics. New York, Plenum Press, 1985, pp 383–396.

25 Al-Musallam A: Revision of the black *Aspergillus* species; thesis, Rijksuniversiteit, Urecht, The Netherlands, 1980.

26 Kusters-van Someren M, Samson RA, Visser J: The use of RFLP analysis in classification of the *Aspergillus niger* aggregate. Curr Genet 1991;19:21–26.

27 Christensen M: The *Aspergillus ochraceus* group: Two new species from western soil and a synoptic key. Mycologia 1982;74:210–225.

28 Christensen M, States JS: *Aspergillus nidulans* group: *Aspergillus navahoensis*, and a revised synoptic key. Mycologia 1982;74:226–235.

29 Horie Y: Ascospore ornamentation and its application to the taxonomic re-evaluation in *Emericella*. Trans Mycol Soc Jpn 1980;21:483–493.

30 Klich MA: Morphological studies of *Aspergillus* section *Versicolores* and related species. Mycologia 1993;85:100–107.

31 Christensen M: A synoptic key and evaluation of species in the *Aspergillus flavus* group. Mycologia 1981;72:1056–1084.

32 Klich MA, Pitt JI: Differentiation of *Aspergillus flavus* from *A. parasiticus* and other closely related species. Trans Br Mycol Soc 1988;91:99–108.

33 Frisvad JC, Samson RA: Chemotaxonomy and morphology of *Aspergillus fumigatus* and related taxa; in Samson RA, Pitt JI (eds): Modern Concepts in *Penicillium* and *Aspergillus* Classification. New York, Plenum Press, 1990, pp 201–208.

34 Geiser DM, Frisvad JC, Taylor JW: Evolutionary relationships in *Aspergillus* section *Fumigati* inferred from partial β-tubulin and hydrophobin DNA sequences. Mycologia 1998;90:831–845.

35 Peterson SW: Phylogenetic analysis of *Aspergillus* sections *Cremei* and *Wentii*, based on ribosomal DNA sequences. Mycol Res 1995;99:1349–1355.

36 Pitt JI, Samson RA: Taxonomy of *Aspergillus* section *Restricti*; in Samson RA, Pitt JI (eds): Modern Concepts in *Penicillium* and *Aspergillus* Classification. New York, Plenum Press, 1990, pp 249–258.

37 Gams W, Christensen M, Onions AH, Pitt, JI, Samson RA: Infrageneric taxa of *Aspergillus*; in Samson RA, Pitt JI (eds): Advances in *Penicillium* and *Aspergillus* Systematics. New York, Plenum Press, 1985, pp 55–62.

38 Horie Y, Miyaji M, Nishimura K, Franco MF, Coelho KLR: New and interesting species of *Emericella* from Brazilian soil. Mycoscience 1996;37:137–144.

39 Greuter W, et al: International Code of Botanical Nomenclature adopted by the 14th Int Bot Congr, Berlin, 1987.

40 Samson RA, Gams W: Typification of the species of *Aspergillus* and associated teleomorphs; in Samson RA, Pitt JI (eds): Advances in *Penicillium* and *Aspergillus* Systematics. New York, Plenum Press, 1986, pp 31–54.

41 Pitt JI: Recent developments in the study of *Penicillium* and *Aspergillus* systematics. J Appl Bact 1989;(symp suppl):37S–45S.

42 Kozakiewicz A, Frisvad JC, Hawksworth DL, Pitt JI, Samson RA, Stolk AC: Proposal for nomina specifica conservanda and rejicienda in *Aspergillus* and *Penicillium*. Taxon 1992;41:109–113.

43 Pitt JI, Samson RA: Species names in current use in the *Trichocomaceae* (Fungi, *Eurotiales*). Regnum Vegetabile 1993;128:13–57.

44 Samson RA, Hoekstra ES, Frisvad JC, Filtenborg O: Introduction to Food-Borne Fungi, ed 5. Baarn, Centraalbureau voor Schimmelcultures, 1996, pp 1–322.

45 Tzean SS, Chen JL, Liou GY, Chen CC, Hsu WH: *Aspergillus* and related teleomorphs of Taiwan. Mycological Monograph No 1. Taiwan, Culture Collection and Research Center, 1990.

46 Pitt JI, Hocking AD: Fungi and Food Spoilage. London, Chapman & Hall, 1997.

47 Klich MA, Samson RA: *Aspergillus* reference cultures. International Commisson of *Penicillium* and *Aspergillus* 1996, pp 1–32.

48 Schimdt A, Wolff MH: Morphological characteristics of *Aspergillus fumigatus* strains isolated from patient samples. Mycoses 1997;40:347–351.

49 Bothast RJ, Fennell DI: A medium for rapid identification and enumeration of *Aspergillus flavus* and related organisms. Mycologia 1974;66:365–369.

50 Pitt JI, Hocking AD, Glenn DR: An improved medium for the detection of *Aspergillus flavus* and *A. parasiticus.* J Appl Bact 1983;54:109–114.

51 Klich MA, Mullaney EJ: Use of bleomycin-containing medium to distinguish *Aspergillus parasiticus* from *A. sojae.* Mycologia 1989;81:159–160.

52 Matsuda H, Kohno S, Maesaki S, Yamada H, Koga H, Tamura M, Kuraishi H, Sugiyama J: Application of ubiquinone systems and electrophoretic comparison of enzymes to identification of clinical isolates of *Aspergillus fumigatus* and several other species of *Aspergillus.* J Clin Microbiol 1992;3:1999–2005.

53 Notermans SHW, Cousin MA, De Ruiter GA, Rombouts FM: Fungal immunotaxonomy; in Frisvad JC, Bridge PD, Arora DK (eds): Chemical Fungal Taxonomy. New York, Marcel Dekker 1998, pp 70–84.

54 Stynen D, Meulemans L, Goris A, Braendlin N, Symons N: Characteristics of a latex agglutination test, based on monoclonal antibodies, for the detection of mould antigen in foods; in Samson RA, Hocking AC, Pitt JI, King ADD (eds): Modern Methods in Food Mycology. Amsterdam, Elsevier, 1992, pp 213–219.

55 Peterson SW: Phylogenetic relationships in *Aspergillus* based upon rDNA sequence analysis; in Samson RA, Pitt JI (eds): Integration of Modern Taxonomic Methods for *Penicillium and Aspergillus* Classification. Dordrecht, Kluwer, 1999.

56 Schmidt A, Schmidt DI: JB Georg W Fresenius and the description of the species *Aspergillus fumigatus*; in Brakhage AA, Jahn B, Schmidt A (eds): *Aspergillus fumigatus:* Biology, Clinical Aspects and Molecular Approaches to Pathogenicity. Contrib Microbiol. Basel, Karger, 1999, vol 2, pp 1–4.

57 Malloch D, Cain RF: The *Trichocomataceae:* Ascomycetes with *Aspergillus Paecilomyces* and *Penicillium* imperfect states. Can J Bot 1972;50:2613–2628.

58 Padhye AA, Godfrey JH, Chandler FW, Peterson SW: Osteomyelitis caused by *Neosartorya pseudofischeri*. J Clin Microbiol 1994;32:2732–2836.

59 Gori S, Pellegrini G, Filipponi F, Della Capana S, Biancofiore G, Mosca F, Lofaro A: Pulmonary aspergillosis caused by *Neosartorya fischeri* (*Aspergillus fischerianus*) in a liver transplant recipient. J Mycol Med 1998;8:105–107.

60 Schmidt A: Diagnostik der Aspergillose. Münch Med Wochenschr 1995;137:607–610.

61 Schmidt A: Pathogenetische Bedeutung von Schimmelpilzen der Gattung *Aspergillus* mit Untersuchungen zur morphologischen Variabilität und in vitro Sensibilität von *Aspergillus fumigatus.* Egelsbach, Hänsel-Hohenhausen Press, 1997.

62 Varsney JL, Sarbhoy AK: A new species of *Aspergillus fumigatus* group and comments upon its synonymy. Mycopathologia 1981;73:89–92.

63 de Vries GA, Cormane RH: A study on the possible relationships between certain morphological and physiological properties of *Aspergillus fumigatus* Fres. and its presence in, or on, human and animal (pulmonary) tissue. Mycol Mycopathol Appl 1969;39:241–253.

64 Leslie LE, Flannigan B, Milner LJR: Morphological studies on clinical isolates of *Aspergillus fumigatus.* J Med Vet Mycol 1988;26:335–341.

65 Debeaupuis JP, Sarfati P, Goris A, Stynen D, Diaquin M, Latgé JP: Exocellular polysaccharides of *Aspergillus fumigatus* and related taxa; in Samson RA, Pitt JI (eds): Modern Concepts in *Penicillium* and *Aspergillus* Classification. New York, Plenum Press, 1990, pp 309–320.

66 Hearn VM, Moutaouakil M, Latgé JP: Analysis of components of *Aspergillus* and *Neosartorya* mycelial preparations by gel electrophoresis and Western blotting techniques; in Samson RA, Pitt, JI (eds): Modern Concepts in *Penicillium* and *Aspergillus* Classification. New York, Plenum Press, 1990, pp 309–320.

67 Polonelli L, Conti A, Campani L, Fanti F: Biotyping of *Aspergillus fumigatus* and related taxa by the yeast killer system; in Samson RA, Pitt JI (eds): Modern Concepts in *Penicillium* and *Aspergillus* Classification. New York, Plenum Press, 1990, pp 309–320.

68 Aho R, Horie Y, Nishimura K, Miyai M: *Aspergillus arvii* spec. nov., a new animal pathogen. Mycoses 1994;37:389–392.

69 Rai JN, Tewari JP, Agarwal SC: The genus *Aspergillus* in India. Lower plants of the Indian subcontinent. Nova Hedwigia 1974;47:487–499.

70 Horie Y, Miyaji M, Nishimura K, Taguchi H, Udagawa S: *Aspergillus fumisynnematus*, a new species from Venezuelan soil. Trans Mycol Soc Jpn 1993;10:3–7.

71 Horie Y: Taxonomic study on *Neosartorya*. J Antibact Antifung Agents 1992;20:27–34.

72 Yaguchi T, Someya A, Udagawa S: A new species of *Neosartorya* from Taiwan soil. Mycoscience 1994;36:309–313.

73 Udagawa S, Tsubouchi H, Toyazaki N: Isolation and identification of *Neosartorya* species from house dust as hazardous indoor pollutants. Mycoscience 1996;37:217–222.

Robert A. Samson, Centraalbureau voor Schimmelcultures, CBS
PO Box 273, NL–3740 AG Baarn (The Netherlands)
Tel. +31 35 54 81234, Fax +31 35 54 16142, E-Mail samson@cbs.knaw.nl

Brakhage AA, Jahn B, Schmidt A (eds): *Aspergillus fumigatus*.
Contrib Microbiol. Basel, Karger, 1999, vol 2, pp 21–43

Pathogenesis and Clinical Presentation of Aspergillosis

R. Rüchel, U. Reichard

Mycology Unit, Department of Medical Microbiology, University of Göttingen, Germany

Invasive aspergillosis is an important opportunistic infection posing a worldwide threat to immunocompromised patients [1]. Whereas an increased rise in invasive *Candida* mycosis in such patients was observed in the eighties, which appears to be receding at present [2, 3], a rise in invasive aspergillosis can be observed these days, which will possibly be remembered as the invasive mycosis of the nineties [4]; our observations support this view (fig. 1). This is the foremost reason for a growing interest in aspergillosis [5–12]. In addition, aspergilli are instrumental in the pathogenesis of certain infections in immunocompetent patients and are relevant agents of allergy. The major diseases with involvement of aspergilli will be reviewed briefly (table 1).

The Infectious Agent

Moulds of the genus *Aspergillus* are asexually reproducing, anamorphic saprobes in the environment which release their infectious propagules, the conidia (spores) from conidiophores into the air. Being hydrophobic on their surface, *Aspergillus* conidia can stay afloat for some time even in moist air, and thus are typical particles of the aerial sediment. The sexually reproducing, teleomorphic counterparts of aspergilli are considered to be noninfectious. Nevertheless, such infections have occasionally been observed in humans and animals, usually involving *Emericella nidulans*, the teleomorph of *A. nidulans* [13, 14]. Several aspergilli, for example, *A. flavus, A. niger, A. terreus* and *A. versicolor* among others are occasionally pathogenic [15, 16]. The foremost

Dedicated to Prof. Dr. Johannes Müller, Emmendingen (Germany).

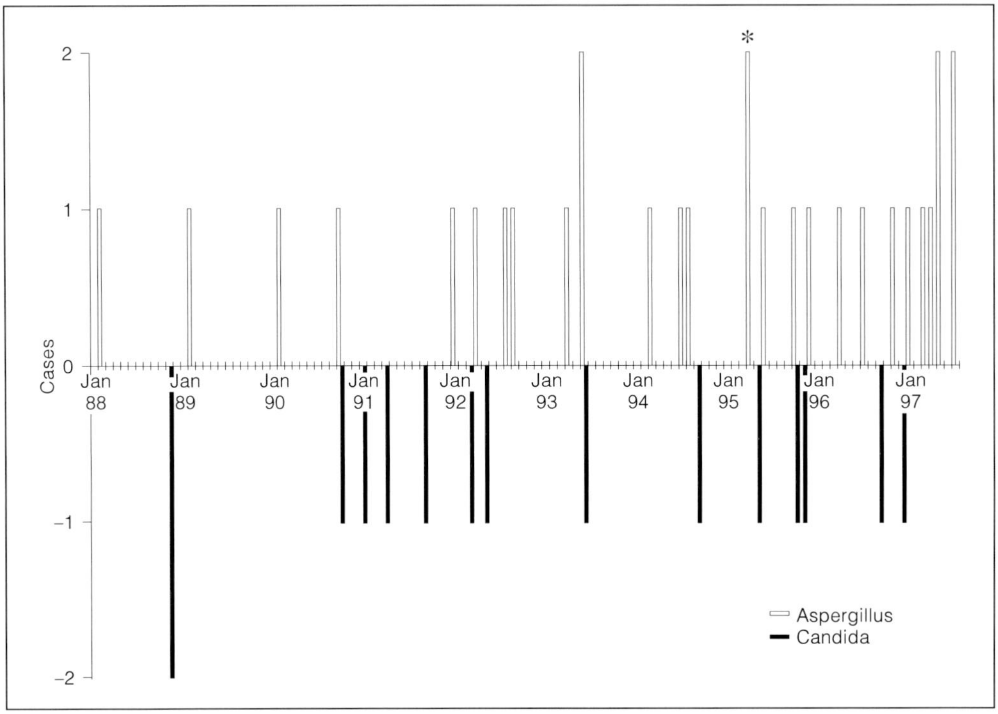

Fig. 1. Cases of proven invasive aspergillosis and invasive candidiasis in the Göttingen University Hospital 1988–1997. *Outbreak afflicting 5 premature infants.

pathogen of the genus, however, is *A. fumigatus* although its conidia account for a surprisingly small fraction of the *Aspergillus* conidia in the environment, for example 0.3% of the aerial *Aspergillus* conidia in a particular cancer hospital [17]. In temperate climates, seasonal concentration shifts of airborne *A. fumigatus* conidia have been observed with minimum values of approximately 5/m^3 in the autumn, and peaks of more than 30 conidia per m^3 in the early spring in an urban environment [18]. With a diameter of 3 μm, conidia of *A. fumigatus* may reach the pulmonar alveoli, but the same may be true for several other aspergilli [16, 19]. The difference between environmental representation and pathogenicity is most evident, when *A. fumigatus* is compared to *A. versicolor* and *A. niger*. Conidia of the latter have been found in large numbers in human dwellings [20]. *A. niger*, in addition, is a typical colonizer of epithelial debris, e.g. of cerumen. However, as an agent of invasive mycosis, both are clearly inferior to *A. fumigatus* [17]. A characteristic, which may be related to the particular pathogenicity of *A. fumigatus* is thermotolerance, since the fungus is able to grow at temperatures of up to 50 °C.

Table 1. Diseases related to aspergilli

Allergic conditions	Infections in immunocompetent patients ( ) = Saprophytic colonization	Infections in immunocompromised patients
Sinusitis	(Sinusitis)	Tracheobronchitis
Extrinsic asthma	(Aspergilloma: nasal, pulmonary)	Chronic necrotizing bronchopneumonia
Alveolitis	(Otitis externa)	Hemorrhagic lung infarction
ABPA	(Cutaneous aspergillosis)	Acute invasive pneumonia
	Invasive pulmonary aspergillosis	Disseminated aspergillosis
	Keratomycosis	Encephalitis, brain abscess
	Endopthalmitis as wound infection	Carditis
	Thoracic/vascular wound infection	Skeletal aspergillosis
	Hematogenous aspergillosis in i.v. drug abusers	Cutaneous aspergillosis
	Chronic disseminated aspergillosis	Thyroid aspergillosis
		Gastrointestinal aspergillosis

Pathogenesis

The pathogenic potential of aspergilli, and particularly of *A. fumigatus*, appears to be based on its surface properties and the release of various extracellular proteins including lytic enzymes or toxins. However, trials to relate pathogenicity to individual factors have failed so far [21, 22]. The dosage of viable conidia taken in, and its relationship to impairment of the respiratory defense are critical, as evidenced by the role of potted plants as reservoirs of aspergilli in the vicinity of neutropenic patients [23, 24], and the potential risk associated with infested dump sites [25]. In the airways the defense against *Aspergillus* conidia is based on clearance by ciliar transport and on phagocytosis by the alveolar macrophages. If the former mechanism is impaired, chronic colonization may be established, as happens in patients with cystic fibrosis [26], whereas invasive infection becomes possible, if the phagocytic potential of the neutrophilic granulocytes is impaired [27], as seen in leukopenic patients. During invasive growth, aspergilli may pervade anatomical borders, e.g. cartilage and the walls of blood vessels, giving rise to hemorrhage or embolic infarction. Also, aspergilli in the airways may release various allergens or toxins [28].

Aspergilli are among the most mycotoxigenic species [29], and the danger of involvement of mycotoxins in the pathogenesis of the *Aspergillus*-related diseases should not be neglected [19]. The mycotoxins are ingested with food, e.g. grain or contaminated dairy products, and can be monitored in blood or urine. Endogenously synthesized gliotoxin has been detected in the tissue of an infected cow [30]. The gliotoxin from *A. fumigatus* acts as an immunodepressant [31], but direct proof of its involvement in human disease has not yet been found. Aflatoxins, originally isolated from toxigenic strains of *A. flavus*, have attracted most interest. The intake of aflatoxin-B with foodstuffs has been related to the incidence of hepatic cancer in patients with chronic viral hepatitis-B [19]. Other conditions which may be related to aflatoxin ingestion are Reye's hepatocerebral syndrome [32]. Ochratoxins are nephrotoxins which were originally detected in grain contaminated by *A. ochraceus*. Chronic exposure to ochratoxin-A by ingestion of contaminated pork may be instrumental in chronic nephritis, furthermore, acute exposure to ochratoxins led to acute renal failure in a patient working in a granary [33]. In this case, the toxin had been generated by fungal conidia. Inhalation of mycotoxigenic conidia with ensuing mycotoxicosis or allergic alveolitis may be considered in connection with early reports on the death of archaeologists who had been opening tombs ('Tutankhamon's curse'). While this is controversial, the following diseases are clearly related to the fungi.

Allergic Diseases Due to Aspergilli

Aspergillus *Asthma*

A rather small fraction of approximately 5% of the asthmatic population suffer from fungal allergy, mostly due to aspergilli but to other more common allergens at the same time [34]. Inhalation of fungal conidia may trigger exacerbation of the disease. Typically, the disease is mediated by IgE; precipitins of the IgG class are missing. However, as many as a quarter of the patients suffering from extrinsic asthma probably have specific precipitins and a positive skin test for *Aspergillus* antigens [35].

Extrinsic Allergic Alveolitis (EAA)

Extrinsic allergic alveolitis due to aspergilli mostly occurs in nonallergic patients after heavy exposure to conidia. It resembles 'organic dust diseases' such as farmer's lung. Symptoms include malaise, fever and dyspnea. Typically, the disease occurs a few hours after intake of aspergilli [34]. Chest X-ray may reveal reticular markings (honeycombing) due to lymphocytic infiltrates. Serum precipitins against *Aspergillus* antigens can be detected and Arthus' reaction can be demonstrated. Depending on the frequency of exposure, pulmonary fibrosis may develop [36]. The treatment is based on corticosteroids and avoidance of the allergens.

Allergic Bronchopulmonary Aspergillosis (ABPA)

ABPA is associated with asthma, it displays a mixture of type I, III and IV allergic reactions, and eosinophilia. Pathogenesis of ABPA is based on conidia of *A. fumigatus* being trapped in the viscous bronchial secretions and releasing allergens and toxic molecules [37, 38]. Patients experience bronchospasms and develop immediate type skin response as a type I (IgE) reaction to fungal antigens. Precipitating IgG antibodies are instrumental in a type III immune complex reaction, and type IV delayed hypersensitivity may develop due to infiltration by mononuclear cells. Patients produce a brownish or blood-streaked purulent sputum, and mucous plugs with eosinophils and viable *Aspergillus* elements. Variable titers of total IgE may serve as a marker of disease activity. Chest roentgenograms may initially reveal fleeting shadows, and atelectasis in the late stage of the disease [34]. Mucoid impaction, preferentially at bifurcations of segmental bronchi, may cause fibrosis, bronchiectasis, and proximal resorption atelectasis [36]. Recurrent episodes of ABPA tend to afflict the same areas, mostly the upper lobes, and lead to chronic obstructive lung disease. ABPA is treated essentially by administration of corticosteroids [39].

Aspergillus Infections in Patients without Evident Immunodeficiency

Aspergilloses in virtually immunocompetent patients are rare diseases [15, 39]. Usually, such mycoses develop in places of tissue damage and become established by saprophytic colonization. They mostly take a slow course and present with formation of granuloma. Important underlying conditions of the lungs are tuberculosis, sarcoidosis, solid tumors, bronchiektasis and related disorders, as well as thoracic trauma. In recent years, invasive aspergillosis has been observed after thoracic or cardiac surgery, e.g. with valvular endocarditis and pseudo-aneurysm. In such cases, also *A. terreus* has repeatedly been identified as causative agent [40] apart from *A. fumigatus*. Outbreaks among patients on intensive care, suggesting a particular environmental source, have been observed, whereby dust generated from indoor or outdoor construction work is particularly suspect [1, 41]. However, endogenous infection from preexisting subclinical pulmonary foci has likewise been experienced (fig. 2). In any case, the search for obscured immunodeficiency is recommended [42].

Cutaneous Aspergillosis

Infection of cutis by aspergilli requires general or localized immunodeficiency, such as the destruction of the protective epidermis in burn wounds. Cutaneous infection by aspergilli in association with foreign bodies has been

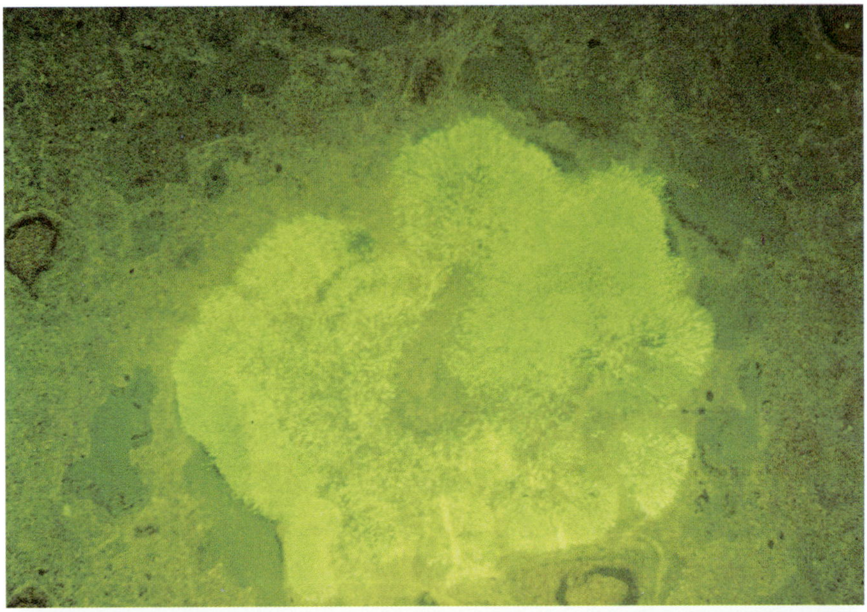

2a

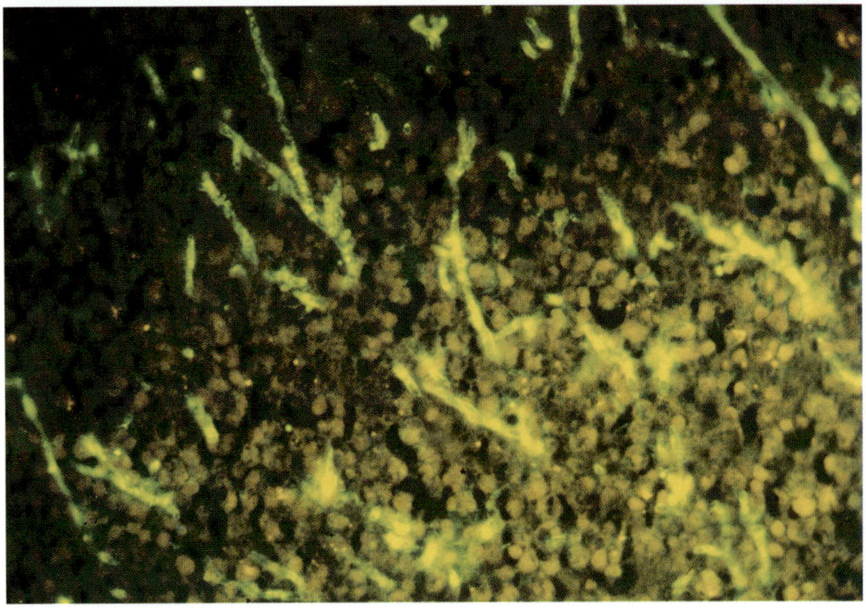

2b

Fig. 2. a Small aspergilloma in the lung of a 24-year-old patient who succumbed to encephalitis following chest trauma. ×60. *b* Disseminated aspergillosis of the brain in the same patient, indirect immunofluorescence using polyclonal antibodies against *Aspergillus* cell wall antigens. Approx. ×400.

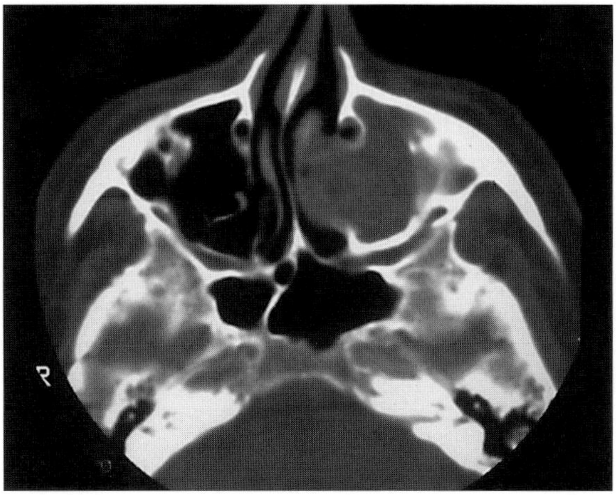

Fig. 3. Aspergilloma of the paranasal sinus of a 69-year-old patient, prior to surgery. CT scan.

observed in compromised and virtually noncompromised individuals, e.g. at insertion sites of intravenous devices and along implants such as tunnelled catheters [43]; its pathogenesis is reminiscent of mycetoma. Aspergillosis on burn wounds has eventually been lethal in half of these patients, suggesting that immunocompromising conditions may have developed in the course of the disease [39]. The afflicted skin usually displays central hemorrhagic necrosis. In the terminal stage of disseminated aspergillosis, a multitude of cutaneous foci may rapidly develop due to embolism.

In premature infants, intimate contact of skin with *Aspergillus* conidia may cause cutaneous infection with ulceration and subsequent invasive diseases.

Sinusitis

Aspergillus-related sinusitis is the most common aspergillosis of the upper airways. It may occur either as an allergic disease with laboratory findings related to asthma, or as acute sinusitis with superficial infection [44]. In some instances, colonization of the paranasal sinus chronifies and gives rise to the formation of structures resembling fungus ball, with deviation of the nasal septum. Pathogenesis of such disease may be related to dental procedures involving use of zinc [45]. If the immune defense of such patients weakens due to age or underlying compromising disease (e.g. diabetes), 'chronic invasive aspergillosis' may ensue, with orbital phlegmon, blindness, thrombosis, and invasion of the base of the brain (fig. 3) [46]. Destruction of adjacent bone, e.g. the palate, or cartilage is common, since the fungus may pervade natural barriers of the host.

Otitis Externa

Mycosis of the outer ear has been observed in humid climates. The superficial infection develops from colonized detritus (cerumen) and is due to *A. niger* in the majority of the cases [39]. Fungal elements may be detected in the discharge from the afflicted ear. As in the sinuses, aspergilloma may develop. If such a structure has access to air, conidiophores may develop (fig. 4).

Invasive Pulmonary Aspergillosis

In patients with severely damaged lungs, invasive infection may establish even without underlying immunodeficiency. Therefore, although rarely, patients with chronic obstructive pulmonary disease may succumb to *Aspergillus* pneumonia [47]. The same appears to be true for chronic alcoholics.

Aspergilloma

Fungus ball is a classic entity in medical mycology. It refers to solid masses of *Aspergillus* mycelia which are glued together by proteins – possibly antibodies – and which usually develop in preexisting cavities. Aspergilloma mostly occurs in the upper lobes of the lungs in tuberculous cavities. Radiographic demonstration of an air crescent or a mobile mass within a cavity are suggestive of the disease. Due to penetration of blood vessels, the aspergilloma patient may experience hemoptysis. If there is a history of severe hemorrhage, the aspergilloma should be treated surgically under antimycotic protection [48, 49]. Detection of fungal elements in the recovered granulomatous mass may require immunofluorescence or molecular genetic means, as morphology of the fungus may present in an atypical fashion. Conservative treatment of aspergilloma may involve intracavitary instillation of amphotericin B. Rarely, aspergilloma has been observed in the urinary tract. Diabetes mellitus appears to be a risk factor for this manifestation of mycosis [39].

Pleuritis and Empyema

Aspergillus pleuritis and empyema are complications of late stage tuberculosis and may develop in the aftermath of thoracic surgery. Sutures of braided

Fig. 4. Aspergilloma of the auditory channel: Hematoxylin-eosin stained section showing conidiophores with biseriate phialides (>), conidia (<), and numerous hyphae. Approx. ×1000.

Fig. 5. Aspergillus endophthalmitis in a 70-year-old patient after corneal grafting. Infiltrated vitreous was removed and stained directly with an optical brightener (Blankophor-P). *A. fumigatus* was grown in parallel. Ca. ×400.

Fig. 6. Subcutaneous node in a 42-year-old patient suffering from chronic disseminated aspergillosis in the aftermath of acute myeloic leukemia [from Kern et al., Ann Haematol, in press]. A biopsy grew *A. fumigatus.*

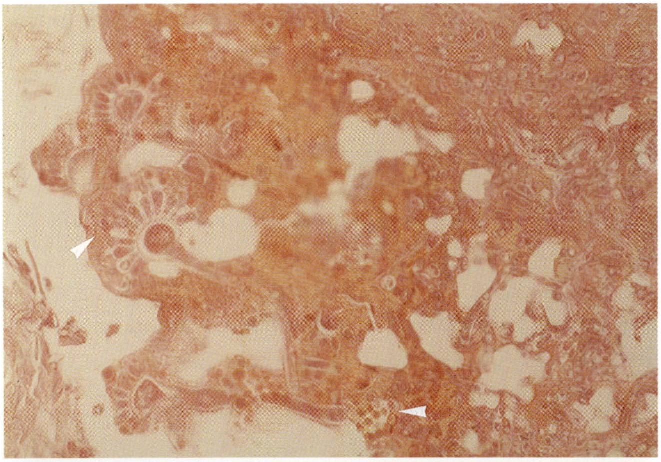

4

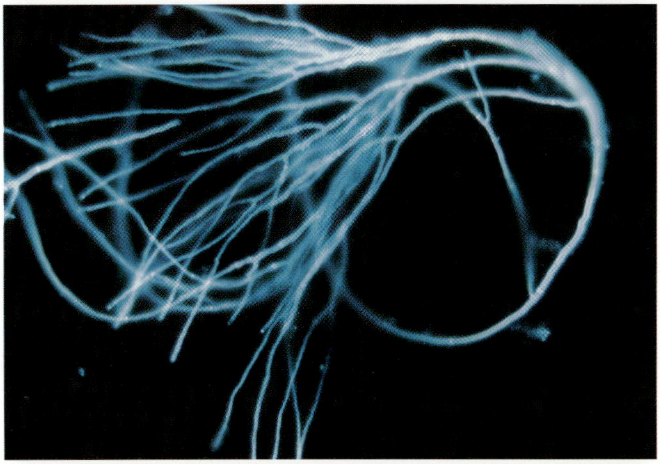

5

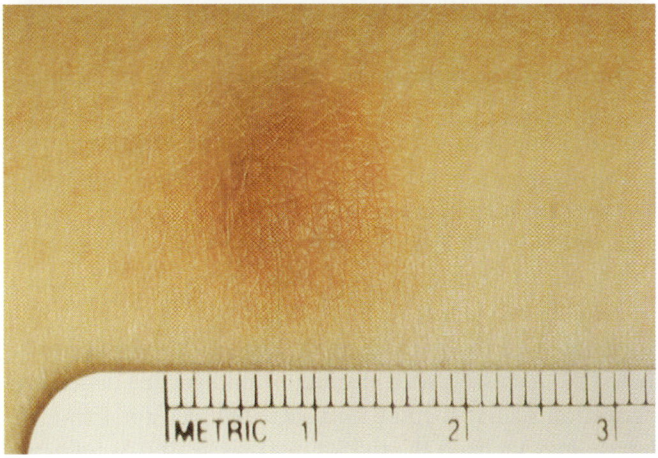

6

silk have been claimed to provide favorable conditions for the establishment of mycosis at these sites [39].

Infections of the Eye

Aspergilli and *Fusarium* species are major agents of fungal keratitis, a disease which is more prevalent in rural areas and tropical climates. It may also afflict wearers of contact lenses, and may lead to corneal ulcer [50].

Aspergillus endophthalmitis in contrast, is a rare disease which may originate from hematogenous spread in immunocompromised patients, or which may establish in immunocompetent patients after ocular trauma or surgery due to contamination (fig. 5). In any case *A. fumigatus* and *A. flavus* are prevalent among the corresponding clinical isolates.

Chronic Disseminated Aspergillosis

Chronic disseminated aspergillosis has been recently proposed as a term for chronic multifocal mycosis in seemingly immunocompetent patients after antineoplastic chemotherapy and in whom the infection became manifest only after return of the peripheral leukocytes [51]. With virtually normalized white blood cell counts, such patients developed mycotic foci, preferentially presenting as subcutaneous granuloma (fig. 6) and destructive foci in the skeleton. Most likely, the infection had been acquired during neutropenia but was suppressed but not eradicated at that time by empiric treatment with amphotericin B. As repeated antineoplastic chemotherapy is likely to result in discrete immunosuppression beyond the end of leukopenia [52], such patients in fact may be considered to be immunocompromised.

Aspergillosis in the Immunocompromised Patient

The major groups of immunocompromised patients threatened by aspergillosis are the thoroughly neutropenic patients, bone marrow transplant patients, and patients with solid organ transplants [53–55]. In fact, 'aspergillosis is perhaps the most greatly feared infectious complication in bone marrow transplant patients, as mortality approaches 100%' [56]. As usual, aspergillosis in these patients will afflict the airways but may take an invasive course, leading to infiltration of the parenchyma, penetration of anatomical borders, and invasion of blood vessels with ensuing hematogenous dissemination. However, the selection of the chemotherapeutic regimen for treatment of the underlying disease may increase the risk of mycosis as well as influence

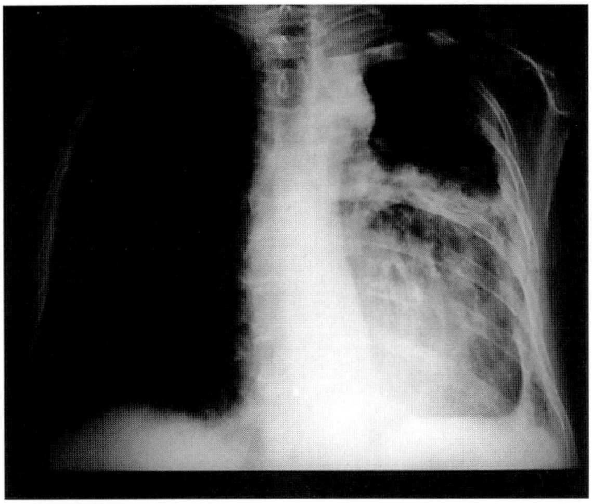

Fig. 7. Chronic necrotizing bronchopneumonia: Chest X-ray of a 65-year-old patient, who yielded numerous respiratory isolates of *A. fumigatus*.

its course [57]. In most such cases aspergillosis could be related to a nosocomial orgin [58].

Tracheobronchitis and Chronic Necrotizing Bronchopneumonia

In patients with moderate immunodeficiency – e.g. diabetes mellitus of low-dose corticosteroid therapy – and particularly in concomitant lung disease such as sarcodiosis, tracheobronchitis with colonization of the airways by *A. fumigatus*, and ensuing ulceration of mucosa may be observed. This condition may chronify and develop into chronic necrotizing bronchopneumonia, a disease which predominantly afflicts patients beyond the age of forty [59]. The patients show symptoms of respiratory disease for at least six months, leukocytosis, specific precipitins of the IgG class, and viable *A. fumigatus* may be recovered from respiratory secretions. Radiography will reveal infiltrates and parenchymal necrosis. About every other patient will develop cavities with aspergilloma (fig. 7). Cavities in an early stage (diameter < 1 cm) can only be detected with certainty by computed tomography. Therapy of necrotizing bronchopulmonary aspergillosis requires surgical measures and amphotericin B treatment.

Hemorrhagic Infarction

Focal aspergillosis of the lung in moderately immunocompromised patients with fungal invasion of arteries may lead to hemorrhagic infarction, which is a typical manifestation of aspergillosis. In chest X-rays, a wedge-like

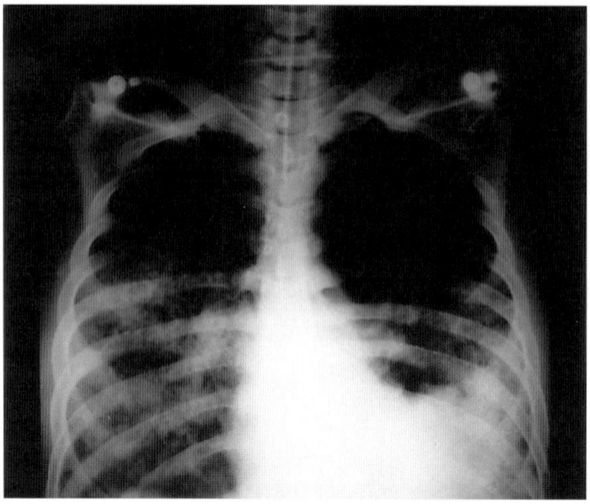

Fig. 8. Lethal acute invasive *A. fumigatus* pneumonia in a 32-year-old chronic alcoholic. Note the bilateral coin lesions in the lower lobes.

shadow as in lung embolism may be observed, with a pleural base. The lesion may display an air crescent sign as in aspergilloma [36].

Invasive Pneumonia

Acute invasive *Aspergillus* pneumonia is possibly the most dreaded infectious disease in severely immunocompromised patients, particularly in those undergoing remission induction chemotherapy for hematological neoplasia. It may occur in patients on long-term high-dose corticosteroids, in neutropenic patients, in organ transplant patients, and occasionally in chronic alcoholics. Particularly during severe neutropenia (<500 neutrophils/mm^3), the disease is usually fatal. Fever, which resists antibiotic therapy, and respiratory failure in a high-risk patient may be due to *Aspergillus* pneumonia. Accompanying chest pain is a sign of pleural involvement. Coin-sized lesions on chest X-rays, often observed firstly in the lower lobes of the lung, are suggestive of the onset of disease (fig. 8). However, with low leukocyte counts pulmonary infiltrates will be poorly developed, leading to underestimation of the extent of the lesions. In computerized chest tomograms, predominantly ill-defined small nodules and consolidation were detected [60], also crescent and halo signs (due to lesser density at the margin) have been associated with mycotic lesions. By magnetic resonance, target images may be demonstrated in later stages of the disease related to a hemorrhagic margin of the lesion and central necrosis [36]. However, this may apply to moderately immunodeficient patients

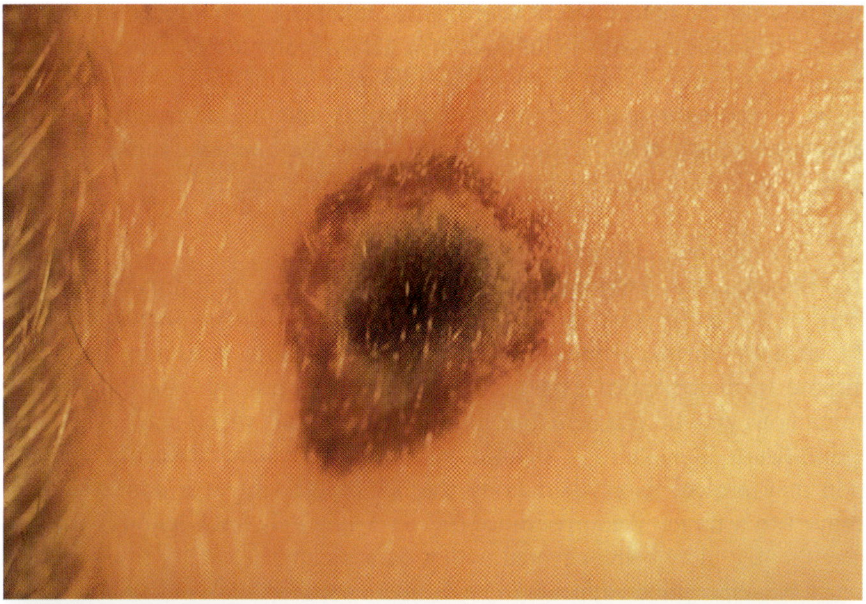

9

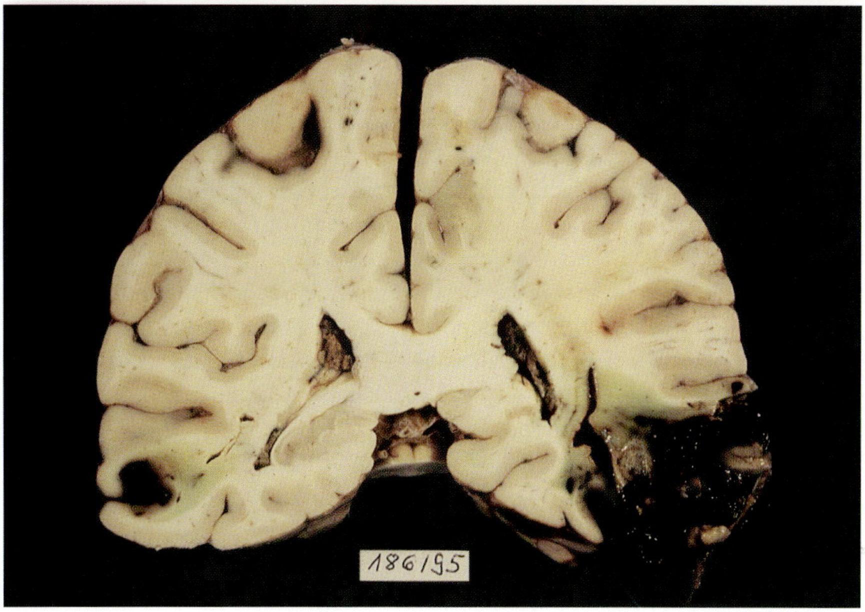

10

Fig. 9. Target lesion on the skin of a 30-year-old patient who developed lethal disseminated aspergillosis during aplastic anemia. *A. fumigatus* was grown from the lesion.

Fig. 10. Cerebral hemorrhage in a 54-year-old patient who succumbed to disseminated aspergillosis after thallium intoxication. Histology of the lesions suggested and culture confirmed *A. fumigatus*. Peripheral leukocytes amounted to 40,000/μl!

and may not be detectable in thoroughly leukopenic patients, in whom the disease will take an overwhelming course. Life-threatening hemoptysis in *Aspergillus* pneumonia may occur in the late stage of the disease and appears to be related to vasculotropism of the fungus and the destructive activity of returning leukocytes. With early-onset high-dose amphotericin B therapy, invasive pulmonary aspergillosis may be treated successfully [61].

Disseminated Aspergillosis

Disseminated, systemic aspergillosis will emanate in approximately one third of the neutropenic patients suffering from *Aspergillus* pneumonia [39]. At virtually normal white blood cell counts, thallium poisoning may also dispose to disseminated aspergillosis. Adjacent organs may be involved by continuous growth, but hematogenous spread prevails which may involve the brain, heart, liver, kidneys, thyroid, spleen and the skin. Metastases on the skin may present as erythematous plaques of the size of a coin and may convert to a target-like appearance (fig. 9). Ultimately, such lesions will turn necrotic, the ulcer being covered with a dark eschar.

CNS-aspergillosis appears to be the main manifestation of hematogenous spread [39, 62]. Signs of encephalitis or stroke in a neutropenic patient suggest aspergillosis which may lead to cerebral hemorrhage and necrosis due to obliteration of blood vessels (fig. 10). In moderately immunodeficient patients, cerebral aspergillosis may present as abscess. *Aspergillus* meningitis, in contrast, has rarely been observed.

The gastrointestinal tract is reported to present the second most prominent target of dissemination, particularly with affliction of the esophagus and the bowel [39, 63].

In the heart, hematogenous aspergillosis may afflict both the endocardium, giving rise to large thrombotic deposits, and to focal myocarditis (fig.11). Likewise, pericarditis and pancarditis have been observed [40].

Multifocal dissemination is the typical pattern of dissemination in the liver, kidneys, spleen and thyroid (fig. 12).

The skeleton may be afflicted by contiguous infection or via hematogenous spread of aspergillosis. In high-risk patients, tumorous foci in the ribs, or spondylodisciitis [64] may become apparent only after the end of induced bone marrow aplasia [51], as prophylactic antimycotic treatment during aplasia becomes more popular among hematologists (fig. 13).

Aspergillosis in HIV Patients

In the late stages of AIDS, patients may experience pulmonary or even disseminated aspergillosis [65, 66]. Corticosteroid therapy or concomitant neutropenia may be instrumental in such cases.

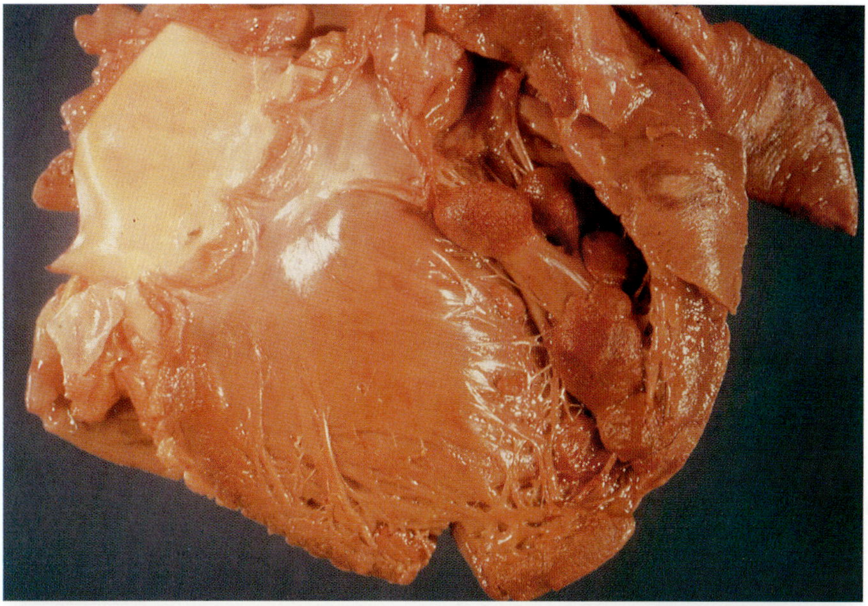

11

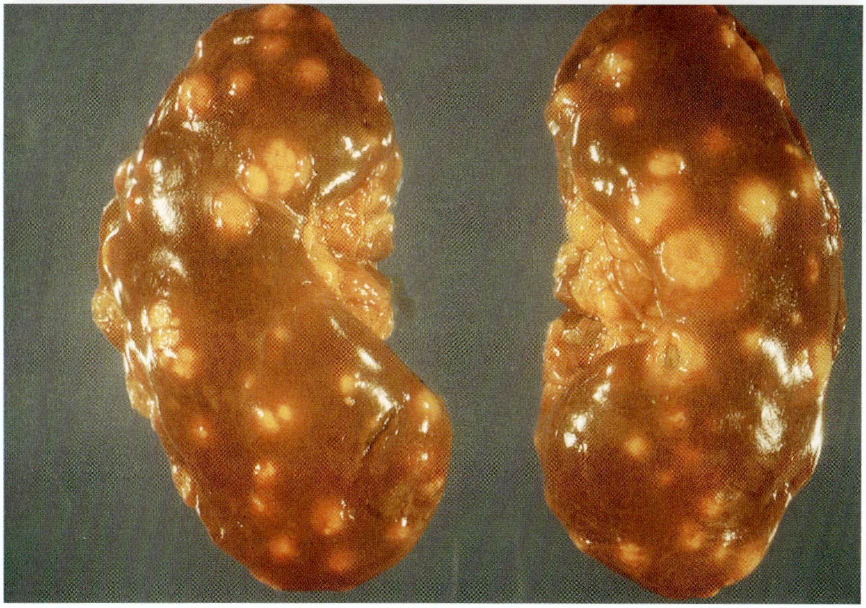

12

Fig. 11. Heart of the same patient as in fig. 9, showing large mycotic vegetations in the right ventricle, from which *A. fumigatus* was grown.

Fig. 12. Renal manifestation of disseminated *A. fumigatus* mycosis (same patient as in fig. 10).

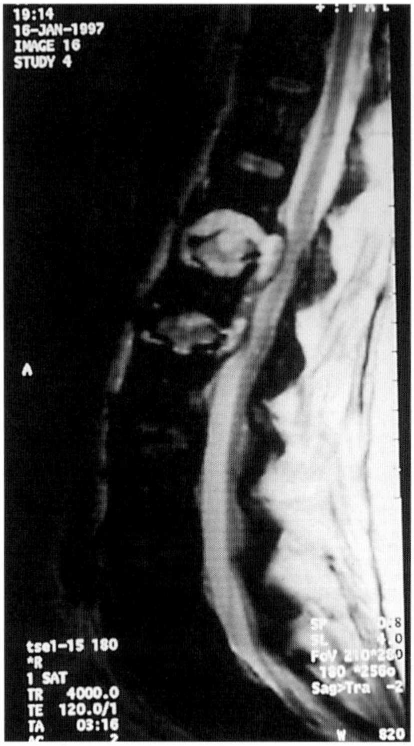

Fig. 13. Spondylodisciitis in a 57-year-old patient who suffered from chronic myeloic leukaemia and developed back pain 6 weeks *after* the end of induced aplasia. *A. fumigatus* was grown from a biopsy. MRI.

Diagnosis

Diagnosis of aspergillosis is established beyond doubt by culture of the mould from infected tissue, whereas blood culture is, for unknown reasons, ineffective, and relatively few cases of *Aspergillus* fungemia have been communicated [67]. A single positive culture from respiratory secretions should not be ignored in immunocompromised patients [68] and, in our experience, should always warrant an immediate alert, irrespective of the assumed immune status of the patient. Monitoring of specific antibodies may be helpful in allergic diseases and in the case of aspergillosis in immunocompetent patients. However, it has proved to be of little help in the high-risk patients, as progress of the disease is often too rapid to allow a suggestive rise of the titer values, and such patients may not be able to mount a regular antibody response due to immunodeficiency [69], an observation which we have confirmed. Earlier attempts at serologic antigen testing had repeatedly met with disappointing results [70], however, an ELISA for improved detection of circulating galactomannan (Platelia®, Sanofi-Pasteur) may be helpful. In our own hands with

215 suspect patients tested in 1997, a positive predictive value of approximately 0.6 could be deduced (11 truly positives versus 7 falsely positives). Sensitivity and specificity of the test have yet to be established [71, 72], since the values approximating 100%, recently cited [73] are, in our experience, for patients not coming exclusively from hematology, too optimistic. For practical purposes, it is advisable, to store an early serum sample from patients, expected to experience bone marrow aplasia, since, if suspicion of aspergillosis arises, testing of paired sera is more reliable. Bronchoalveolar lavage (BAL) is a popular procedure for recovery of fungal elements from respiratory secretions, but it needs further improvement for a safe diagnosis [74]. Microscopic detection of aspergilli in smears or tissue sections requires staining. Gram staining is unreliable, but PAS or Grocott staining is useful, though time-consuming. In tissue sections, hematoxylin-eosin staining is adequate. Fungal elements in bronchial secretions can be reliably detected by fluorescence microscopy using a suitable filter kit, following staining of the specimen with an optical brightener (fig. 5). Since the brighteners (e.g. Blankophor-P; Bayer AG, Leverkusen, Germany) are stable in 20% KOH, maceration of specimens and fluorescent staining of liberated fungal elements can be performed simultaneously, rendering this procedure applicable to any type of clinical specimen. Such preparations should be examined as wet mounts [75]. The demonstration of hyaline, septate hyphae with dichotomous branching, mostly at the angle of 45°, is suggestive of aspergillosis. However, histology may be deceiving, since other hyalohyphomycetes like *Fusarium* or *Scedosporium* species cannot be distinguished with certainty by these criteria, and immunohistological differentiation is laborious [76]. *Aspergillus* rarely produces its typical conidiophores in the infected organs, as this requires free access to oxygen. A wrong diagnosis can be detrimental, as *Fusarium* or *Scedosporium* isolates may resist treatment with amphotericin B, which is effective against aspergilli. Tissue with infiltrates of *A. niger* may display crystals of calcium oxalate, which can be detected in polarized light [77]. Attempts for amplification of *Aspergillus* DNA by polymerase chain reaction (PCR) from clinical specimens have yielded promising results, particularly when primers were employed, which allow the detection of many fungal species, and with subsequent differentiation, e.g. by sequencing of the amplimer [78]. However, the procedure still appears to be too demanding to allow its use in the diagnostic routine of an average mycological laboratory, as fungi require particular measures for the liberation of DNA. With respiratory secretions, PCR is particularly prone to produce false-positive results due to contamination and irrelevant colonization [79]. Therefore, at present, molecular diagnostic techniques do not improve the diagnosis of fungal infection of the respiratory tract [80]. In spite of these diagnostic options, aspergillosis is still often detected only by necropsy [81].

Prophylaxis and Therapy of Aspergillosis: Remarks

The therapy of *Aspergillus*-related diseases has been described by Ellis [this vol.]. However, a few remarks may be warranted: Prophylaxis for aspergillosis in high-risk patients firstly requires reduction of the fungal burden in the enviroment, e.g. by removal of potted plants [23, 24], and protection from dust developing at construction sites. Ultimately, protection by laminar airflow and high-efficiency particulate (HEPA) air filtering may be indicated. Antimycotic prophylaxis for aspergillosis has not definitely been established: Itraconazole is only fungistatic and does not directly reach the airways, while inhalation of amphotericin B aerosols has yielded disappointing results [82–85]. The therapy of invasive aspergillosis is based on intravenous amphotericin B up to 1.25 mg/kg/day [86]. According to our own experience, abrupt onset of highly dosed intravenous amphotericin B may be crucial, allowing for peak blood levels of 1.5 μg/ml [15]. Amphotericin B-lipid formulas appear to be at least as efficient as conventional amphotericin B and less nephrotoxic even at elevated dosage [87, 88]. However, a distribution study of liposomal amphotericin B with relatively low concentrations in the lungs [89], suggested that related preparations may be less favorable for the treatment of pulmonary mycosis. The high cost of amphotericin B-lipid formulations prohibits their use in empirical therapy, but a combined diagnostic approach including antigen monitoring and radiological imaging may provide for a selection of those patients, who deserve such treatment [73]. The supplementation of amphotericin B with the growth factor GM-CSF may shorten neutropenia, but has been blamed for causing hemoptysis [90]. Itraconazole appears to present an alternative therapeutic option in moderately or nonimmunocompromised patients [91]. Its benefit, however, is limited, as it acts in a fungistatic fashion and is restricted to oral application. Fluconazole has occasionally been tried in aspergillosis. However, as aspergilli are usually resistant, breakthrough aspergillosis with a lethal outcome has occurred [92]. Recently, a combined surgical and antimycotic approach to invasive pulmonary aspergillosis has been recommended [93]. While this approach appears to be promising in moderately immunosuppressed patients, it is difficult to conceive of such an approach in the peracute aspergillosis of thoroughly immunodeficient patients.

Conclusions

Depite the continuing discovery of 'new' fungal pathogens [94], aspergilli remain the second most important group of agents of invasive mycosis (aspergillosis) and may even take the lead over the *Candida* species. Aspergillosis is possibly the most dreaded infectious complication for leucopenic or transplant patients. In spite of some progress being made with the diagnosis of aspergillosis, the prognosis of patients suffering from the invasive type of the disease has not improved much during the last decade, particularly, as no really new antimycotic agents have become available, and nonconventional delivery forms of amphotericin B or biological response modifiers have not brought about a major breakthrough in the therapy of the infection. Hence, the awareness of the clinician, early diagnosis, and empiric amphotericin B therapy are still the main options for a successful outcome of the battle against the disease. Aspergillosis in many cases, including moderately immunodeficient patients, is a nosocomial disease requiring effective measures to reduce the risk of exposure to infectious conidia by identification and elimination of potential sources of infection from the patient's environment. Furthermore, aspergilli remain an important source of allergic diseases. The role of *Aspergillus* in medical mycology was highlighted in rhymes by Barranco [95].

Acknowledgment

Several photographs were kindly provided by colleagues from the Göttingen University Hospital: W. Brück (Department of Neuropathology), J. Dörner (Department of Orthopedics); W. Kern, M. Kneba and T. Waller (Department of Medicine); and B. Sattler and A. Schauer (Department of Pathology).

References

1 Fridkin SK, Jarvis WR: Epidemiology of nosocomial fungal infections. Clin Microbiol Rev 1996; 9:499–511.

2 Debusk CH, Daoud R, Thirumoorthi MC, Wilson FM, Khatib R: Candidemia: Current epidemiologic characteristics and a long-term follow-up of the survivors. Scand J Infect Dis 1994;26:697–703.

3 Fisher-Hoch SP, Hutwagner L: Opportunistic candidiasis: An epidemic of the 1980s. Clin Infect Dis 1995;21:897–904.

4 Wald A, Leisenring W, Burik van JA, Bowden RA: Epidemiology of *Aspergillus* infections in a large cohort of patients undergoing bone marrow transplantation. J Infect Dis 1997;175: 1459–1466.

5 Musial CE, Cockerill FR, Roberts GD: Fungal infections of the immunocompromised host: Clinical and laboratory aspects. Clin Microbiol Rev 1988;1:349–364.

6 Walsh TJ, Pizzo PA: Nosocomial fungal infections: A classification for hospital-acquired fungal infections and mycoses arising from endogenous flora or reactivation. Ann Rev Microbiol 1988; 42:517–545.

7 Saral R: *Candida* and *Aspergillus* infections in immunocompromised patients: An overview. Rev Infect Dis 1991;13:487–492.

8 Rhodes JC, Jensen HE, Nilius AM, Chitambar CR, Farmer SG, Washburn RG, Steele PE, Amlung TW: *Aspergillus* and aspergillosis. J Med Vet Mycol 1992;30(suppl 1):51–57.

9 Paya CV: Fungal infections in solid-organ transplantation. Clin Infect Dis 1993;16:677–688.

10 Andriole VT: Infections with *Aspergillus* species. Clin Infect Dis 1993;17:S481–S486.

11 Armstrong D: Overview of invasive fungal infections and clinical presentation. Baillière's Clin Infect Dis 1995;2:17–24.
12 Schwartz S, Thiel E: Clinical presentation of invasive aspergillosis. Mycoses 1997;40(suppl 2):21–24.
13 Mitchell RG, Chaplin AJ, Mackenzie DWR: *Emericella nidulans* in maxillary sinus fungal mass. J Med Vet Mycol 1987;25:339–341.
14 Guillot J, Collobert C, Gueho E, Mialot M, Lagarde E: *Emericella nidulans* as an agent of guttural pouch mycosis in a horse. J Med Vet Mycol 1997;35:433–435.
15 Kwon-Chung KJ, Bennett JE: Medical Mycology. Philadelphia, Lea & Febiger, 1992, pp 201–247.
16 De Hoog GS, Guarro J: Atlas of Clinical Fungi. Baarn, Centraalbureau voor Schimmelcultures, 1995, pp 136–139.
17 Schmitt HJ, Blevins A, Sobeck K, Armstrong D: *Aspergillus* species from hospital air and from patients. Mycoses 1991;33:539–541.
18 Mullins J, Harvey R, Seaton A: Sources and incidence of airborne *Aspergillus fumigatus* (Fres). Clin Allergy 1976;6:209–217.
19 Pitt JI: The current role of *Aspergillus* and *Penicillium* in human and animal health. J Med Vet Mycol 1994;32(suppl 1):17–32.
20 Beguin H, Nolard N: Mould biodiversity in homes. I. Air and surface analysis of 130 dwellings. Aerobiologia 1994;10:157–166.
21 Bouchara JP, Tronchin G, Larcher G, Chabasse D: The search for virulence determinants in *Aspergillus fumigatus*. Trends Microbiol 1995;3:327–330.
22 Hogan LH, Klein B, Levitz SM: Virulence factors of medically important fungi. Clin Microbiol Rev 1996;9:469–488.
23 Summerbell RC, Krajden S, Kane J: Potted plants in hospitals as reservoirs of pathogenic fungi. Mycopathologia 1989;106:13–22.
24 Staib F: Pathogenic fungi in human dwellings. Mycoses 1992;35:289–292.
25 Kramer MN, Kurup VP, Fink JN: Allergic bronchopulmonary aspergillosis from a contaminated dump site. Am Rev Respir Dis 1989;140:1086–1088.
26 Wojnarowski C, Eichler I, Gartner C, Götz M, Renner S, Koller DY, Frischer T: Sensitization to *Aspergillus fumigatus* and lung function in children with cystic fibrosis. Am J Respir Crit Care Med 1997;155:1902–1907.
27 Schaffner A, Douglas H, Braude A: Selective protection against conidia by mononuclear and against mycelia by polymorphonuclear phagocytes in resistance to *Aspergillus*. J Clin Invest 1982;69:617–619.
28 Latgé JP, Paris S: The fungal spore, reservoir of allergens; in Cole GT, Hoch HC (eds): The Fungal Spore and Disease Initiation in Plants and Animals. New York, Plenum Press, 1991, pp 379–401.
29 Samson RA: Mycotoxins: A mycologist's perspective. J Med Vet Mycol 1992;30(suppl 1):9–18.
30 Bauer J, Gareis M, Bott A, Gedek B: Isolation of a mycotoxin (gliotoxin) from a bovine udder infected with *Aspergillus fumigatus*. J Med Vet Mycol 1989;27:45–50.
31 Sutton P, Waring P, Müllbacher A: Exacerbation of invasive aspergillosis by the immunosuppressive fungal metabolite, gliotoxin. Immunol Cell Biol 1996;74:318–322.
32 Hogan GR, Ryan NJ, Hayes AW: Aflatoxin B and Reye's syndrome. Lancet 1978;i:561.
33 Di Paolo N, Guarnieri A, Loi F, Sacchi G, Mangiarotti AM, Di Paolo M: Acute renal failure from inhalation of mycotoxins. Nephron 1993;64:621–625.
34 Beaumont F: Clinical manifestations of pulmonary *Aspergillus* infections. Mycoses 1988;31(suppl 2): 15–20.
35 Bromley IMJ, Donaldson K: Binding of *Aspergillus fumigatus* spores to lung epithelial cells and basement membrane proteins: Relevance to the asthmatic lung. Thorax 1996;51:1203–1209.
36 Rau WS: *Aspergillus* infections of the lung: Radiological signs. Mycoses 1997;40(suppl 2):25–32.
37 Slavin RG, Gottlieb CC, Avioli LV: Allergic bronchopulmonary aspergillosis. Arch Intern Med 1986;146:1799–1801.
38 Robinson BWS, Venaille TJ, Mendis AHW, McAleer R: Allergens as proteases: An *Aspergillus fumigatus* proteinase directly induces human epithelial cell detachment. J Allergy Clin Immunol 1990;86:726–731.
39 Bodey GP, Vartivarian S: Aspergillosis. Eur J Clin Microbiol Infect Dis 1989;8:413–437.
40 Kuttin ES: Fungal infections of the cardiovascular system: A review. Mycoses 1997;40:3–24.

41 Humphreys H, Johnson EM, Warnock DW, Willatts SM, Winter RJ, Speller DCE: An outbreak of aspergillosis in a general ITU. J Hosp Infect 1991;18:167–177.
42 Kelly JK, Pinto AR, Whitelaw WA, Rorstad OP, Bowen TJ, Matheson DS: Fatal *Aspergillus* pneumonia in chronic granulomatous disease. Am J Clin Pathol 1986;86:235–240.
43 Allo MD, Miller J, Townsend T, Tan C: Primary cutaneous aspergillosis associated with Hickman intravenous catheters. N Engl J Med 1987;317:1105–1108.
44 De Shazo RD, Chapin K, Swain RE: Fungal sinusitis. N Engl J Med 1997;337:254–259.
45 Willinger B, Beck-Mannagetta J, Hirschl AM, Makristathis A, Rotter ML: Influence of zinc oxide on *Aspergillus* species: A possible cause of local noninvasive aspergillosis of the maxillary sinus. Mycoses 1995;39(suppl 1):20–25.
46 Karim M, Alam M, Shah AA, Ahmed R, Sheikh H: Chronic invasive aspergillosis in apparently immunocompetent hosts. Clin Infect Dis 1997;24:723–733.
47 Pittet D, Huguenin T, Dharan S, Sztajzel-Boissard J, Ducel G, Thorens JB, Auckenthaler R, Chevrolet JC: Unusual cause of lethal pulmonary aspergillosis in patients with chronic obstructive pulmonary disease. Am J Respir Crit Care Med 1996;154:541–544.
48 Glimp RA, Bayer AS: Pulmonary aspergilloma. Arch Intern Med 1983;143:303–308.
49 Kauffman CA: Quandary about treatment of aspergillomas persists. Lancet 1996;347:1640.
50 Thomas PA: Mycotic keratitis: An underestimated mycosis. J Med Vet Mycol 1994;32:235–256.
51 Rüchel R, Kern W: Chronic systemic aspergillosis. Mycoses 1997;40:1.
52 Sable CA, Donowitz GR: Infections in bone marrow transplant recipients. Clin Infect Dis 1994; 18:273–284.
53 Guiot HFL, Fibbe WE, van't Wout JW: Risk factors for fungal infection in patients with malignant hematologic disorders: Implications for empirical therapy and prophylaxis. Clin Infect Dis 1994; 18:525–532.
54 Guillemain R, Lavarde V, Amrein C, Chevalier P, Guinvarc'h A, Glotz D: Invasive aspergillosis after transplantation. Transplant Proc 1995;27:1307–1309.
55 Keating MR, Guerrero MA, Daly RC, Walker RC, Davies SF: Transmission of invasive aspergillosis from a subclinically infected donor to three different organ transplant recipients. Chest 1996;109: 1119–1124.
56 LaRocco MT, Burgert SJ: Infection in the bone marrow transplant recipient and role of the microbiology laboratory in clinical transplantation. Clin Microbiol Rev 1997;10:277–297.
57 Bow EJ, Loewen R, Cheang MS, Schacter B: Invasive fungal disease in adults undergoing remission-induction therapy for acute myeloid leukemia: The pathogenetic role of the antileukemic regimen. Clin Infect Dis 1995;21:361–369.
58 Walsh TJ, Dixon DM: Nosocomial aspergillosis: Environmental microbiology, hospital epidemiology diagnosis and treatment. Eur J Epidemiol 1989;5:131–142.
59 Binder RE, Faling LJ, Pugatch RD, Mahasaen C, Snider GL: Chronic necrotizing pulmonary aspergillosis: A discrete clinical entity. Medicine 1982;61:109–124.
60 Logan PM, Primack SL, Miller RR, Müller NL: Invasive aspergillosis of the airways: Radiographic, CT, and pathologic findings. Radiology 1994;193:383–388.
61 Ruchlemer R, Yinnon AM, Hershko C: Changes in the natural history of invasive pulmonary aspergillosis in neutropenic leukemic patients. Israel J Med Sci 1996;32:1089–1092.
62 Salaki JS, Louria DB, Chmel H: Fungal and yeast infections of the central nervous system. Medicine 1984;63:108–132.
63 Catalano L, Picardi M, Anzivino D, Insabato L, Notaro R, Rotoli B: Small bowel infarction by *Aspergillus*. Haematologica 1997;82:182–183.
64 Scully RE, Mark EJ, McNeely WF, McNeely BU: A 65-year-old man with vertebral and disk lesion after a pulmonary operation for *Aspergillus* infection. N Engl J Med 1991;324:754–763.
65 Lortholary O, Meyohas MC, Dupont B, Cadranel J, Salmon-Ceron D, Peyramond D, Simonin D: Invasive aspergillosis in patients with acquired immunodeficiency syndrome: Report of 33 cases. Am J Med 1993;95:177–187.
66 Tumbarello M, Tacconelli E, Pagano L, Ortu La Barbera EO, Morace G, Cauda R, Leone G, Ortona L: Comparative analysis of prognostic indicators of aspergillosis in haemotological malignancies and HIV infection. J Infect 1997;34:55–60.

67 Duthie R, Denning DW: *Aspergillus* fungemia: Report of two cases and review. Clin Infect Dis 1996;20:598–605.
68 Kusne S, Torre-Cisneros J, Manez R, Irish W, Martin M, Fung J, Simmons RL, Starzl TE: Factors associated with invasive lung aspergillosis and the significance of positive *Aspergillus* culture after liver transplantation. J Infect Dis 1992;166:1379–1383.
69 Young RC, Bennett JE: Invasive aspergillosis: Absence of detectable antibody response. Am Rev Respir Dis 1971;104:710–716.
70 Hopwood V, Johnson EM, Cornish JM, Foot ABM, Evans EGV, Warnock DW: Use of the Pastorex *Aspergillus* antigen latex agglutination test for the diagnosis of invasive aspergillosis. J Clin Pathol 1995;48:210–213.
71 Sulahian A, Tabouret M, Ribaud P, Sarfati J, Gluckman E, Latgé JP, Derouin F: Comparison of an enzyme immunoassay and latex agglutination test for detection of galactomannan in the diagnosis of invasive aspergillosis. Eur J Clin Microbiol Infect Dis 1996;15:139–145.
72 Bretagne S, Costa JM, Marmorat-Khuong A, Poron F, Cordonnier C, Vidaud M, Fleury-Feith J: Serum *Aspergillus* galactomannan antigen testing by sandwich ELISA: Practical use in neutropenic patients. J Infect 1997;35:7–15.
73 Severens JL, Donnelly JP, Meis JFGM, de Vries Robbé PF, de Pauw BE, Verweij PE: Two strategies for managing invasive aspergillosis: A decision analysis. Clin Infect Dis 1997;25:1148–1154.
74 Delvenne P, Arrese JE, Thiry A, Borlee-Hermans G, Pierard GE, Boniver J: Detection of cytomegalovirus, *Pneumocystis carinii*, and *Aspergillus* species in bronchoalveolar lavage fluid. Am J Clin Pathol 1993;100:414–418.
75 Rüchel R, Margraf S: Rapid microscopical diagnosis of deep-seated mycoses following maceration of fresh specimens and staining with optical brighteners. Mycoses 1993;36:239–242.
76 Kaufman L; Standard PG, Jalbert M, Kraft D: Immunohistologic identification of *Aspergillus* spp and other hyaline fungi by using polyclonal fluorescent antibodies. J Clin Microbiol 1997;35:2206–2209.
77 Ghio AJ, Petersheim DS, Roggli VL, Piantadosi CA: Pulmonary oxalate deposition associated with *Aspergillus niger* infection. Am Rev Respir Dis 1992;145:1499–1502.
78 Einsele H, Hebart H, Roller G, Löffler J, Rothenhofer I, Müller CA, Bowden RA, van Burik JA, Engelhard D, Kanz L, Schumacher U: Detection and identification of fungal pathogens in blood by using molecular probes. J Clin Microbiol 1997;35:1353–1360.
79 Bretagne S, Costa JM, Marmorat-Khuong A, Poron F, Cordonnier C, Vidaud M, Fleury-Feith J: Detection of *Aspergillus* species DNA in bronchoalveolar lavage samples by competitive PCR. J Clin Microbiol 1995;33:1164–1168.
80 Ieven M, Goossens H: Relevance of nucleic acid amplification techniques for diagnosis of respiratory tract infections in the clinical laboratory. Clin Microbiol Rev 1997;10:242–256.
81 Bodey G, Bueltmann B, Duguid W, Gibbs D, Hanak H, Hotchi M, Mall G, Martino P, Meunier F, Milliken S, Naoe S, Okudaira M, Scevola D, van't Wout J: Fungal infections in cancer patients: An international autopsy survey. Eur J Clin Microbiol Infect Dis 1992;11:99–109.
82 Beyer J, Schwartz S, Heinemann V, Siegert W: Strategies in prevention of invasive pulmonary aspergillosis in immunosuppressed or neutropenic patients. Antimicrob Agents Chemother 1994;38:911–917.
83 Schuler US, Haag C: Prophylaxis of fungal infections. Mycoses 1997;40(suppl 2):41–44.
84 Lortholary O, Dupont B: Antifungal prophylaxis during neutropenia and immunodeficiency. Clin Microbiol Rev 1997;10:477–504.
85 Erjavec Z, Woolthuis GMH, de Vries-Hospers HG, Sluiter WJ, Daenen SMGJ, de Pauw B, Halie MR: Tolerance and efficacy of amphotericin B inhalations for prevention of invasive pulmonary aspergillosis in haematological patients. Eur J Clin Microbiol Infect Dis 1997;16:364–368.
86 Armstrong D: Treatment of opportunistic fungal infections. Clin Infect Dis 1993;16:1–9.
87 Prentice HG, Hann IM, Herbrecht R, Aoun M, Kvaloy S, Catovsky D, Pinkerton CR, Schey SA, Jacobs F, Oakhill A, Stevens RF, Darbyshire PJ, Gibson BES: A randomized comparison of liposomal versus conventional amphotericin B for the treatment of pyrexia of unknown origin in neutropenic patients. Br J Haematol 1997;98:711–718.

88 White MH, Anaissie EJ, Kusne S, Wingard JR, Hiemenz JW, Cantor A, Gurwith M, Du Mond C, Mamelok RD, Bowden RA: Amphotericin B colloidal dispersion vs amphotericin B as therapy for invasive aspergillosis. Clin Infect Dis 1997;24:635–642.

89 Van Etten EWM, Otte-Lambillion M, van Vianen W, ten Kate MT, Bakker-Woudenberg IAJM: Biodistribution of liposomal amphotericin B (AmBisome) and amphotericin B-desoxycholate (Fungizone) in uninfected immunocompetent mice and leucopenic mice infected with *Candida albicans*. J Antimicrob Chemother 1995;35:509–519.

90 Groll A, Renz S, Gerein V, Schwabe D, Katschan G, Schneider M, Hübner K, Kornhuber B: Fatal haemoptysis associated with invasive pulmonary aspergillosis treated with high-dose amphotericin B and granulocyte-macrophage colony-stimulating factor (GM-CSF). Mycoses 1992;35:67–75.

91 Denning DW, Lee JY, Hostetler JS, Pappas P, Kauffman CA, Dewsnup DH, Galgiani JN, Graybill JR, Sugar AM, Catanzaro A, Gallis H, Perfect JR, Dockery B, Dismukes WE, Stevens DA: NIAID mycoses study group multicenter trial of oral itraconazole therapy for invasive aspergillosis. Am J Med 1994;97:135–144.

92 Meis JF, Donelly JP, Hoogkamp-Korstanje JA, de Pauw BE: *Aspergillus fumigatus* pneumonia in neutropenic patients during therapy with fluconazole for infection due to *Candida* species. Clin Infect Dis 1993;16:734–735.

93 Bernard A, Caillot D, Couaillier JF, Casasnovas O, Guy H, Favre JP: Surgical management of invasive pulmonary aspergillosis in neutropenic patients. Ann Thorac Surg 1997;64:1441–1447.

94 Perfect JR, Schell WA: The new fungal opportunists are coming. Clin Infect Dis 1996;22:S112–S118.

95 Barranco CP: *Aspergillus* and its place in medical mycology. J Med Vet Mycol 1994;33:477–479.

Prof. Dr. R. Rüchel, Hygiene Institut, Universität Göttingen,
Kreuzbergring 57, D–37075 Göttingen (Germany)
Tel. +49 551 395 855, Fax +49 551 395 860

Brakhage AA, Jahn B, Schmidt A (eds): *Aspergillus fumigatus*.
Contrib Microbiol. Basel, Karger, 1999, vol 2, pp 44–56

Epidemiology and Molecular Basis of the Involvement of *Aspergillus fumigatus* in Allergic Diseases

Reto Crameri

Swiss Institute of Allergy and Asthma Research (SIAF), Davos, Switzerland

Members of the genus *Aspergillus* can be pathogenic, saprophytic or allergenic agents, depending on the immune status of the host. For example, *A. fumigatus* may be a pathogen, causing invasive aspergillosis in immunocompromised patients [1], a saprophyte, causing aspergillomata in patients who are immunologically normal but who have lung damage due to diseases such as tuberculosis or sarcoidosis [2], or an allergen for atopic individuals suffering from asthma or cystic fibrosis patients [3, 4]. However, *A. fumigatus* rarely affects healthy individuals in spite of a generally occurring exposure demonstrated by the isolation of fungi from the sputum of normal individuals [5]. Therefore, not only the virulence of the fungus but also the immune status of the host play a predominant role in these opportunistic mycoses [6]. Each of the *A. fumigatus*-related pulmonary complications, ranging from benign colonisation of the lung to life-threatening diseases such as invasive aspergillosis or allergic bronchopulmonary aspergillosis (ABPA) [7], are characterised by specific immune responses in the host [2]. Beside the recent progress achieved in understanding the pathophysiological mechanisms of defence against fungal infections showing the importance of polymorphonuclear cells in the phagocytosis of important pathogenic yeasts [8] and in a damage of hyphae of *A. fumigatus* [9], problems related to a reliable diagnosis remain to be solved [10]. Demonstration of circulating antigens and antibodies raised against fungal components are considered as important criteria for the diagnosis of *Aspergillus*-related diseases [2, 11–13]. Hallmarks in the diagnosis of a clinical suspicion of *A. fumigatus*-sensitisation are the radioallergosorbent test (RAST) and prick or intradermal skin tests with fungal extracts, corroborated by various serologic methods [14]. However, the clinical relevance of an *A. fumi-*

gatus sensitisation determined by skin test or by the demonstration of allergen-specific IgE is only reliable if defined antigen/allergen preparations are applied [2, 15]. The currently available fungal extracts are not satisfactory and need to be improved for dependable results. *A. fumigatus* extracts are complex mixtures containing up to 200 different proteins and glycoproteins [16, 17] and many low-molecular-weight compounds [18, 19]. The fungus has been reported to be able to produce more than 40 IgE-binding components [20] which generate complex immunopatterns when extracts are assayed with sera from allergic individuals in Western blots [20, 21]. Problems encountered in the standardisation of *A. fumigatus* extracts are due to different patterns of allergens produced by different clinical isolates [22], batch to batch variations [23], time dependent liberation of IgE-binding components during growth [24], instability of the extract due to protease content [25], and culture conditions used to grow the fungus [26]. The lack of defined *A. fumigatus* antigen/allergen preparations of high purity has hampered the analysis of *A. fumigatus*-associated diseases at cellular and molecular level [17]. In this regard, cloning, characterisation and production of recombinant *A. fumigatus* allergens can substantially contribute to provision of the long awaited standardised reagents for clinical and basic research applications [15, 27].

Environmental Distribution and Epidemiology of *A. fumigatus*

Although the genus *Aspergillus* includes about 132 species and 18 variants thereof [28], only a few species are recognised as opportunistic pathogens for man. *A. fumigatus* is the etiological agent isolated in about 80% of the *Aspergillus*-related opportunistic mycoses in humans [29], however, other species such as *A. terreus, A. flavus, A. nidulans, A. oryzae*, and *A. niger* have been occasionally associated with aspergillosis [2]. *Aspergillus* conidia are ubiquitously distributed in man's environment and have been recovered from the upper atmosphere, the snow in the Antarctic, and the sand and winds over the Sahara desert [30]. An analysis recorded that the genus *Aspergillus* was between first and fifteenth in order of incidence of airborne moulds and varied in frequency between 0.1 and 22% of the total air spores [31]. Studies have been carried out in hospitals to determine the presence of the pathogen. *A. fumigatus* was found in the ventilation system of operation rooms, mattress dust, in-between false ceilings and plants at the bedside [32]. This may be of particular importance because in hospitals patients undergoing antineoplastic chemotherapy, organ transplantation or antibiotic therapy are at risk of developing opportunistic mycoses [1]. Recoveries of *Aspergillus* colonies within domestic and commercial interiors commonly exceed those obtained from free air [31, 33]. However, opinions on the importance of the number of *Aspergillus* conidia in the immediate environment of patients vary, and systematic aerobiological studies in the microenvironments of patients are not available [2].

Detailed epidemiological investigation may not be necessary when the causative agent can be directly correlated with a particular source or with a specific disease. In terms of development of adverse allergic reactions, *A. fumigatus* mainly affect patients suffering from asthma [3] or cystic fibrosis (CF) [4]. In western Europe about 20% of patients with asthma have a positive skin test to fungal extracts [34–36], while the prevalence of *A. fumigatus* sensitisation in CF patients has been reported to occur in as much as 31–59% of the CF population [37–39], suggesting a high prevalence of sensitisation to the fungus in these patients. There is more disagreement about the incidence of ABPA in both groups of patients [36, 39]. ABPA, a hypersensitivity disease of the lung resulting in a chronic inflammatory response of proximal bronchi and distal air space associated with an array of immune responses to *A. fumigatus*, has been reported to occur with an incidence of between 0.9 and 23% in CF patients and between 7 and 14% in asthmatic individuals sensitised to *A. fumigatus* [3, 4, 38, 40–43]. The wide ranges of prevalence are probably due to a variable interpretation of clinical and laboratory criteria being used to define ABPA [44, 45]. The diagnosis of ABPA is complicated by a number of characteristics shared with underlying diseases. The clinical and laboratory findings associated with ABPA often encountered in CF and asthmatic patients without ABPA include sensitisation to *A. fumigatus*, positive precipitins and elevated specific serum IgE and IgG antibodies to *A. fumigatus* extracts, elevated total serum IgE as well as pulmonary infiltrates [44, 45]. Centrally located bronchiectasis, one of the classical criteria for the diagnosis of ABPA in allergic asthmatics [46], occurs in CF patients without ABPA at high frequency [44]. Therefore, serological findings should strongly contribute to confirm or exclude ABPA suspected on clinical symptoms [2, 36, 39]. However, the reliability of diagnostic procedures such as skin tests or serologic investigations mainly depends on the quality of the antigen preparations used [47, 48]. With regard to *A. fumigatus*, even standardised commercial preparations are not satisfactory [49, 50] emphasising the need for standardised reagents to reliably diagnose *A. fumigatus* sensitisation [15].

Fungal Extracts and Diagnosis of *A. fumigatus*-Related Diseases

The diagnosis of *A. fumigatus*-related diseases and problems encountered are reviewed in a separate chapter of this book and do not need to be discussed in detail here (see Kappe and Rimek; Rüchel and Reichard; this volume).

However, the reliability of a laboratory diagnosis largely depends on the quality of the reagents used. Therefore, the American Academy of Allergy, Asthma and Immunology [47] recently recommended the use of standardised extracts – where available – for the diagnosis and treatment of allergic disease. However, in spite of significant progress in allergen standardisation achieved in recent years, standardised mould extracts approved by international organisations are still not available [47, 48]. Commercially available *A. fumigatus* extracts for skin testing or serology are not standardised in sensu stricto and therefore extracts from different commercial sources do generate discrepant results as recently shown in a study involving skin prick tests in patients with CF [49]. Moreover, the use of different *A. fumigatus* extracts for in vivo and in vitro testing may account for the reported discrepancies between skin test results and serological results [51]. Problems related to the standardisation of *A. fumigatus* extracts have been largely discussed and are mainly related to strain variabilities [22], batch to batch variations [23], time-dependent liberation of IgE-binding proteins during growth [24], extract instabilities [25], culture and harvesting conditions [26], and extraction procedures used [52]. Even standardised extracts will hardly allow a discrimination between ABPA and *A. fumigatus* sensitisation, because the immunological background of both diseases is related to polyclonal IgE-mediated responses to fungal exposure [2, 15]. However, Western blot analyses with fungal extracts using sera of individuals suffering from ABPA or *A. fumigatus* sensitisation showed significant differences in the patterns of IgE-binding proteins recognised by the two groups of patients [20]. These observations indicate that a specific sensitisation to single *A. fumigatus* allergens can occur and prompted us to clone and characterise *A. fumigatus* allergens [15, 27] to study the differential immune responses to single allergens in the different *A. fumigatus*-related diseases [36, 39].

Cloning *A. fumigatus* Allergens: The Filamentous Phage Approach

A. fumigatus is a complex allergenic source estimated to be able to produce more than 40 IgE-binding proteins [20]. Although classical techniques have proven to be successful for the determination of the primary structure of allergens [53, 54], cloning of several components from complex allergenic systems like *A. fumigatus* is labor-intensive. A rational cloning approach for cDNAs encoding many different proteins sharing common binding properties like allergens, and to clone cDNAs at all, would imply selective enrichment of specific clones expressing proteins able to interact with the corresponding natural ligand [55–59]. Screening of cDNA expression libraries constructed in bacteriophage lambda vectors or plasmid systems requires immobilisation

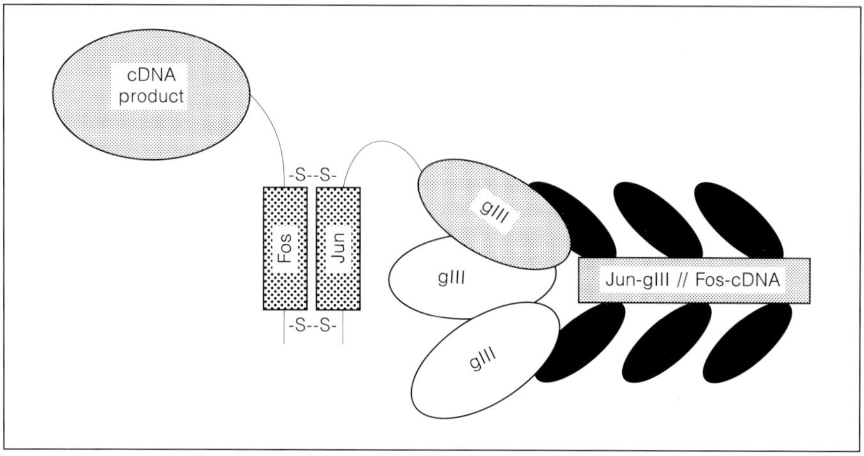

Fig. 1. Schematic representation of the physical linkage between cDNA products displayed on phage surface and genetic information required for their production achieved by Jun/Fos-heterodimerisation. From Crameri [15].

of the library to a solid phase, and detection of positive clones is achieved by interaction with ligands displaying affinity to the required proteins [60]. The immobilisation of libraries to a solid phase hampers, however, selective enrichment of clones expressing proteins of interest [60]. A prerequisite for the specific enrichment of clones from cDNA expression libraries is the physical linkage between gene product and its cDNA, a requirement not fulfilled by most available cloning systems. Filamentous phage surface display systems overcome this problem because cloned cDNA and expressed gene product become packaged as an entity through physical linkage of gene product displayed on phage surface and replication functions (fig. 1) [58, 61]. However, such systems allowing selection of genes from cDNA libraries were not available because translational stop codons present at the 3′-end of nontranslated regions of eukaryotic mRNA prevent construction of fusion proteins N-terminal to phage coat gene products, while fusion to the membrane-bound C-terminus would hamper the incorporation of the fusion proteins into the phage coat [59]. We have developed a filamentous phage cloning system, formally termed pJuFo, which allows the display of cDNA products on the phage surface by heterodimerisation of a modified gIII (gene III product) phage coat protein carrying the Jun leucine zipper as N-terminal fusion with cDNA products decorated with the Fos leucine zipper (fig. 1) [55–59]. The most important advantage of the pJuFo cloning system compared to classical cloning systems results from the fact that the phage library can be kept in a liquid phase, where only phages with affinity to the solid phase-coated ligand are

Table 1. Proteins displayed on pJuFo and selected with different ligands

Protein	Function	Source	Ligand	Reference
Asp f 1	allergen	*A. fumigatus* cDNA library	human IgE	55
PhoA	phosphatase	*E. coli*	–	55
β-Actin	structural protein	human cDNA library	HIV reverse transcriptase	57, 59
IgG1, Fc	antibody fragment	human cDNA library	SPA	57, 59
Asp f 3–13	allergens	*A. fumigatus* cDNA library	human IgE	15, 27, 56, 58
Tropomyosin	structural protein	human cDNA library	Jun zipper	57
LHCP	light harvesting proteins	chloroplast cDNA library	antibodies	59
λ-Repressor	DNA-binding protein	λ-phage DNA	λ-phage DNA	57
Cdk's	kinases	human & hamster	cyclin	59
DHFR	reductase	mouse	antibodies	59
Api X	allergens	celery cDNA library	human IgE	Sommergruber-Hoffmann and Crameri, unpubl.
Cop X	allergens	*Coprinus* cDNA library	human IgE	Brander and Crameri, unpubl.

retained on the surface after washing [57, 58]. Phages are biological systems able to autoreplicate in a suitable host. As a direct consequence thereof, desorbed phages can be used to infect *E. coli* in order to amplify phages with affinity for the ligand. Therefore, successive rounds of phage growth and selection allow specific enrichment of phage-displaying proteins with affinity for the ligand [56]. The pJuFo cloning system has been successfully used to enrich phages displaying proteins encoded by cDNAs from different sources including antibody fragments, DNA-binding proteins and several strictly intracellular cyclin dependent kinases (table 1). With regard to cloning allergens from organisms producing complex patterns of IgE-binding proteins, screening of libraries constructed in pJuFo reduces the time required for the selection of candidate clones to a few days [27, 54]. Screening of an *A. fumigatus* cDNA library displayed on the surface on filamentous phage by panning against human IgE from *A. fumigatus*-allergic individuals yielded a vast variety of clones encoding different allergenic proteins [15, 27]. The physical linkage between gene product and its encoding cDNA allows straightforward determination of the primary structure of the protein by nucleotide sequence determination of the cDNA [57, 58]. Moreover, subcloning of the inserts into suitable expression vectors allows the production of recombinant allergens in large amounts [15, 54]. The molecular nature and clinical relevance of the allergens isolated is discussed below.

Table 2. Characterisation of cloned *A. fumigatus* allergens

Name[a]	MW kD	Biological function	IgE-binding[b] properties	Skin test[c] reactivity	References
rAsp f 1	16.86	ribotoxin	++++	+	50, 64, 65
rAsp f 2	34	unknown	++	NT	75
rAsp f 3	18.45	peroxisomal protein	++++	+	73
rAsp f 4	30.04	unknown	++	+	15, 36, 39
rAsp f 5	42.06	metalloprotease	++++	+	15, 27
rAsp f 6	23.03	MnSOD[d]	++	+	66, 67
rAsp f 7	11.63	unknown	++	NT	15, 27
rAsp f 8	11	P2 ribosomal protein	++	NT	Hemann and Crameri, unpubl.
rAsp f 9	33.7	unknown	++++	NT	15, 27
rAsp f 10	34.4	aspartyl protease	++	+	15, 27
rAsp f 11	20	cyclophilin	++++	NT	Hemmann and Crameri, unpubl.
rAsp f 13[e]	19.8	serine protease?	++	NT	Hemann and Crameri, unpubl.

[a] Nomenclature approved by the Allergen Nomenclature Committee [54].

[b] Sera from 60 *A. fumigatus*-sensitised patients were investigated for the presence of allergen-specific IgE [50]. ++++ = Proteins recognised by >50%, ++ = proteins recognised by <50% of the sera tested.

[c] Intradermal skin test reactivity determined in clinical studies. NT = Not tested.

[d] Manganese superoxide dismutase.

[e] Nomenclature not officially approved.

Molecular Nature and Clinical Relevance of Proteins Involved in *A. fumigatus* Allergy

A list of the *A. fumigatus* allergens so far cloned and characterised including known biological functions, molecular weight, clinical tests performed and references is reported in table 2. One of the most important findings is that none of the *A. fumigatus* allergens identified so far with respect to sequence and/or enzymatic function correspond to a previously described nonfungal allergen. However, many of the allergens do occur as homologous allergens in more than one fungal species and could account for the cross-reactivity between moulds [62]. In the case of Asp f 3, cross-reactivity with two peroxisomal proteins of *Candida boidinii* has been demonstrated at the molecular level by comparison of the reactivity of the recombinant proteins [63]. Asp f 1 [64], a potent fungal ribotoxin [65], rAsp f 5, a metalloprotease [15] and rAsp f 10, an aspartic proteinase [15], represent species-specific allergens so far cloned only from fungi of the genus *Aspergillus.* Asp f 6 (manganese superoxide dismutase [66, 67]), Asp f 8 (P2 acidic ribosomal protein) and Asp f 11 (peptidyl-prolyl cis-trans isomerase) belong to the class of phylogenetically highly conserved proteins including human and are the most interesting allergens from the immunological point of view. MnSOD has been shown to be a pan-allergen in patients sensitised to *A. fumigatus* MnSOD [67]. The 23-kD protein shares 51.5% identity and 67.2% similarity to the corresponding human enzyme. Both human and *A. fumigatus* MnSOD coding sequences were expressed as recombinant proteins in *E. coli* and shown to be recognised by IgE antibodies from subjects allergic to the *A. fumigatus* MnSOD. Moreover, both proteins elicited specific immediate type allergic skin reactions in these individuals [66]. Both MnSOD's induced proliferation in peripheral blood mononuclear cells of *A. fumigatus*-allergic subjects who showed specific IgE responses and reacted in skin tests to MnSOD [66]. These observations provide evidence for autoreactivity to the human MnSOD in allergic individuals sensitised to an environmental allergen from *A. fumigatus* who share a high degree of sequence identity to the corresponding human enzyme. Whether the cause of these autoimmune reactions is due to molecular mimicry between conserved T- and B-cell epitopes present on the fungal and on the structurally related human enzyme or by sensitisation to human MnSOD due to an inflammatory process remains to be elucidated [66]. It will be interesting to see if the two allergens, rAsp f 8 and rAsp f 11, which also share a high degree of sequence similarity with the corresponding human proteins will show similar features. However, the responses against a self antigen in individuals sensitised to the *A. fumigatus* MnSOD indicate an autoantigenic T-cell-mediated pathogenesis in chronic inflammatory responses to the fungus growing in the lung of ABPA patients. In many infectious diseases, dominant antibody responses are directed towards microbial antigens with similarity to self proteins. Molecular mimicry at the T-cell level to homologous peptide sequences from human and microbial antigens has been demonstrated in several autoimmune diseases [66].

An additional *A. fumigatus* allergen (Asp f 13) shares a high degree of sequence similarity to a *Coccidioides immitis* antigen of 19 kD [Hemman et al., unpubl.]. *C. immitis* is a soil-borne fungus and causative agent of a human respiratory disease known as coccidioidomycosis [68] which is also an emerging pulmonary infection in patients with AIDS [69]. Again we meet similarities at antigen level between fungi that mainly affect the respiratory tract of immunocompromised individuals. A serological investigation involving patients suffering from aspergillosis, Asp f 13 and the *C. immitis* 19-kD antigen will clarify if humoral cross-reactivity between the two proteins does exist.

An array of other allergenic proteins (Asp f 4, Asp f 7, Asp f 9) and three other not yet fully sequenced allergens fail to show any similarity to already known proteins. These proteins are only characterised by their ability to bind serum IgE from *A. fumigatus*-sensitised individuals. However, they can substantially contribute to improve the diagnosis of ABPA as in the case of rAsp f 4 [36, 39] and to detect sensitisation to *A. fumigatus* in the other cases [15].

A major concern in using recombinant allergens for the diagnosis of allergic diseases is, whether the proteins produced in heterologous hosts are biologically active. The final demonstration that an allergen preparation acts as such in vivo is its ability to elicit type 1 skin reactions in sensitised individuals. In vivo activity needs to be demonstrated specially for recombinant proteins produced in heterologous expression systems which may not be correctly folded. Unfolded allergens are unlikely to be able to provoke mediator release in vivo as demonstrated for the major honeybee venom allergen Api m 1, the soluble phospholipase A_2 [70, 71]. The recombinant allergens Asp f 1, Asp f 3, Asp f 4 and Asp f 6 have been tested in large scale skin studies and demonstrated to be reliable diagnostic reagents [15, 49, 50, 72]. For Asp f 5 and Asp f 10 limited skin test studies showed the allergenicity of the recombinant protein preparations [15, 27]. Moreover, the allergens Asp f 1, Asp f 3, Asp f 4 and Asp f 6 have been immobilised to the Pharmacia ImmunoCAP system and evaluated in a form suitable for routine assessments in a fully automated diagnostic system [73, 74], whereas no discrepancy between skin test and serology was found [15, 50]. These data suggest the possibility of relying in serological data for diagnosis of *A. fumigatus* sensitisation and for discrimination between ABPA and *A. fumigatus* sensitisation [15, 36, 39] improving the sensitivity of the diagnosis on one hand and resulting in a benefit for the patient deriving from the avoidance of skin challenges on the other hand.

Conclusions and Outlook

The development of a phage surface display technology for cDNA cloning allowed rapid progress in the identification and characterisation of allergenic molecules produced by *A. fumigatus*. The fully sequenced allergens of the mould can be subdivided into two classes: proteins with known biochemical function and proteins without sequence similarity with any known proteins. Clinically, all recombinant allergens tested so far are able to elicit strong type 1 skin reactions in sensitised individuals indicative of a biological relevance for the native proteins. Large-scale serological studies showed that responses to at least two allergens (rAsp f 4 and rAsp f 6) are highly specific for sera of patients suffering from ABPA. These two allergens allow a serological discrimination between ABPA and *A. fumigatus* sensitisation with a 100% specificity and 90% sensitivity which was never attained so far. These results show that disease-specific allergens do exist in aspergillosis and open new perspectives for a differential diagnosis of *A. fumigatus*-related pulmonary complications. This important conclusion dictates the screening strategy for the isolation of disease-specific allergens. We are currently evaluating a further panel of *A. fumigatus* allergens representing potential candidates for ABPA-specific allergens and so far investigated only in Western blot analyses. The availability of a wide panel of recombinant *A. fumigatus* allergens will allow selection of sera from individuals suffering from different IgE-mediated *A. fumigatus* complications lacking sensitisation to already cloned allergens. These sera, together with the powerful screening procedure based on biopanning, will facilitate isolation of clones encoding IgE-binding proteins. Production, characterisation and evaluation of the new allergens is likely to allow further improvement of the differential diagnosis of *A. fumigatus*-related allergic diseases substantially contributing to the solution of an old diagnostic problem.

Acknowledgements

I am indebted to Prof. Dr. K. Blaser for his continuous encouragement and to the members of the Swiss Institute of Allergy and Asthma Research (SIAF) for helpful discussions, particularly to S. Hemmann, C. Mayer and U. Appenzeller for practical support. I thank Dr. A. Faith (SIAF) for critical comments on the manuscript. The work was only possible with the intensive clinical support and I am indebted to Dr. G. Menz and Dr. C. Ismail (Hochgebirgsklinik, Davos-Wolfgang, Switzerland), Prof. Dr. M. H. Schöni and Dr. W. H. Nikolaizik (Alpine Kinderklinik, Davos, Switzerland), and Dr. R. Disch (Klinik für Dermatologie und Allergologie, Davos, Switzerland). The work was supported by the Swiss National Science Foundation (Grants 31.50515.97 and 31.39429.93) and by the Ciga-Geigy Jubiläumsstiftung.

References

1 Saral R: *Candida* and *Aspergillus* infections in immunocompromised patients: An overview. Rev Infect Dis 1991;13:487–492.
2 Kurup VP, Kumar A: Immunodiagnosis of apergillosis. Clin Microbiol Rev 1991;4:439–456.
3 Greenberger PA, Patterson R: Allergic bronchopulmonary aspergillosis and the evaluation of the patient with asthma. J Allergy Clin Immunol 1988;81:646–650.
4 Nelson LA, Callerame LM, Schwarz RH: Aspergillosis and atopy in cystic fibrosis. Am Rev Respir Dis 1979;20:863–873.
5 Comstock GW, Palmer LC, Stone RW, Goodman NL: Fungi in the sputum of normal men. Mycopathol Mycol Appl 1974;54:55–63.
6 Romani L, Howard DH: Mechanisms of resistance to fungal infections. Curr Opin Immunol 1995; 7:517–523.

7 Bardana EJ: The clinical spectrum of aspergillosis. 2. Classification and description of saprophytic allergic and invasive variants of human disease. CRC Crit Rev Clin Lab Sci 1980;13:85–159.
8 Lyman CA, Walsh TJ: Phagocytosis of medically important yeasts by polymorphonuclear leukocytes. Infect Immun 1994;62:1489–1493.
9 Roilides E, Uhlig K, Venzon D, Pizzo PA, Walsh TJ: Enhancement of oxidative response and damage caused by human neutrophils to *Aspergillus fumigatus* hyphae by granulocyte colony stimulating factor and gamma interferon. Infect Immun 1993;61:1185–1193.
10 Patterson R, Greenberger, PA, Castile RG, Yee WF, Roberts M: Diagnostic problems in hypersensitivity lung disease. Allergy Proc 1989;10:141–147.
11 Bennet EJ: Rapid diagnosis of candidiasis and aspergillosis. Rev Infect Dis 1987;9:398–402.
12 Rogers TR, Haynes KA, Barnes RA: Value of antigen detection in predicting invasive pulmonary aspergillosis. Lancet 1990;336:1210–1213.
13 Menz G, Ismail C, Crameri R: Die allergische bronchopulmonale Aspergillose. Pneumologie 1996; 50:419–427.
14 Greenberger PA, Patterson R: Allergic bronchopulmonary aspergillosis and the evaluation of the patient with asthma. J Allergy Clin Immunol 1988;81:646–650.
15 Crameri R: Recombinant *Aspergillus fumigatus* allergens: From the nucleotide sequence to clinical applications. Int Arch Allergy Immunol 1998;115:99–114.
16 Piechura JE, Huang CJ, Cohen SH, Kidd JM, Kurup VP, Calvanico NJ: Antigens of *Aspergillus fumigatus*. II. Electrophoretic and clinical studies. Immunol 1983;49:657–665.
17 Slavin RG, Knutsen AP: Purified *Aspergillus* proteins: Going where no one has gone before. J Lab Clin Med 1993;121:80–381.
18 Müllbacher A, Eichner RD: Immunosuppression in vitro by a metabolite of a human pathogenic fungus. Proc Natl Acad Sci USA 1984;81:3835–3837.
19 Williams BP, Nolte H, Dolen WK, Koepke JW, Selner JC: The histamine content of allergen extracts. J Allergy Clin Immunol 1992;89:738–745.
20 Borga Å: Allergens of *Aspergillus fumigatus*; PhD thesis, Karolinska Institute, Stockholm, Repro Print AB, 1990.
21 Leser C, Kauffman HF, Virchow C, Menz G: Specific serum immunopatterns in clinical phases of allergic bronchopulmonary aspergillosis. J Allergy Clin Immunol 1992;90:589–599.
22 Wellenbeck I, Aukrust L, Einarsson R: Antigenic variability of different strains of *Aspergillus fumigatus*. Int Archs Allergy Appl Immunol 1984;73:166–172.
23 Longbottom JL, Austwick PKC: Antigens and allergens of *Aspergillus fumigatus*. I. Characterization by quantitative immunoelectrophoretic techniques. J Allergy Clin Immunol 1986;78:9–17.
24 Kauffman HF, van der Heide S, van der Laan S, Hovenga H, Beaumont F, de Vries K: Standardization of allergenic extracts of *Aspergillus fumigatus:* Liberation of IgE-binding components during cultivation. Int Archs Allergy Appl Immunol 1985;76:168–173.
25 Esch RE: Role of proteases on the stability of allergenic extracts; in Klein R (ed): Regulatory Control and Standardization of Allergen Extracts. Stuttgart, Fischer, 1990, pp 171–177.
26 Kauffman HF, van der Heide S, de Vries K: Antigenic composition of *Aspergillus fumigatus* in relation to the conditions of growth; in Poucard T, Dreborg S (eds): Mould Allergy Workshop. Uppsala, Pharmacia, 1994, pp 43–44.
27 Crameri R, Blaser K: Cloning *Aspergillus fumigatus* allergens by the pJuFo filamentous phage display system. Int Arch Allergy Immunol 1996;110:41–45.
28 Raper KB, Fenell DI: The Genus *Aspergillus*. Baltimore, William & Wilkins, 1965.
29 Al-Doory Y, Wagner GE: Aspergillosis. Springfield, Thomas, 1985.
30 Bardana EJ: The clinical spectrum of aspergillosis. 2. Epidemiology, pathogenicity, infection in animals and immunology of *Aspergillus*. CRC Crit Rev Clin Lab Sci 1980;13:21–84.
31 Solomon WR, Mathews KP: Aerobiology and inhalant allergens; in Middelton E, Reed CE, Ellis EF, Adkinson NF, Yunginger JW (eds): Allergy, Principle and Practice. St Louis, Mosby, 1988, pp 312–358.
32 Prahl P: Reduction of indoor airborne mould spores. Allergy 1992;47:362–365.
33 Solomon WR, Burge HP, Boise JR: Airborne *Aspergillus fumigatus* levels outside and within a large clinical centre. J Allergy Clin Immunol 1978;62:56–60.

34 Hendrick DJ, Davies RJ, D'Souza MF, Pepys J: An analysis of skin prick test reaction in 656 asthmatic patients. Thorax 1975;30:2–8.
35 Schwartz HJ, Citron KM, Chester EH, Kaimal J, Barlow PB, Baum GL, Schuyler MR: A comparison of the prevalence of sensitisation to *Aspergillus* antigens among asthmatics in Cleveland and London. Clin Exp Immunol 1978;62:9–14.
36 Crameri R, Hemmann S, Ismail C, Menz G, Blaser K: Disease-specific recombinant allergens for the diagnosis of allergic bronchopulmonary aspergillosis. Int Immunol, in press.
37 Milla CE, Wielinski CL, Regelmann WE: Clinical significance of the recovery of *Aspergillus* species from the respiratory secretions of cystic fibrosis patients. Pediatr Pulmonol 1996;21:6–10.
38 Valletta EA, Braggion C, Mastella G: Sensitization to *Aspergillus* and allergic bronchopulmonary aspergillosis in a cystic fibrosis population. Ped Asthma Allergy Immunol 1993;7:43–49.
39 Hemmann S, Nikolaizik WH, Schöni MH, Blaser K: Differential IgE recognition of recombinant *Aspergillus fumigatus* allergens by cystic fibrosis patients with allergic bronchopulmonary aspergillosis or *Aspergillus* allergy. Eur J Immunol 1998;28:1155–1160.
40 Nikolaizik WH, Brueton MJ, Warner JO: *Aspergillus* allergy and allergic bronchopulmonary aspergillosis in cystic fibrosis. Pediatr Allergy Immunol 1991;2:83–86.
41 El-Dahr JM, Fink R, Selden R, Arruda KL, Plattsd-Mills TAE, Heymann PW: Development of immune responses to *Aspergillus* at an early age in children with cystic fibrosis. Am J Respir Crit Care Med 1994;150:1513–1518.
42 Basica JE, Graves TS, Baz MN: Allergic bronchopulmonary aspergillosis in corticosteroid dependent asthmatics. J Allergy Clin Immunol 1981;68:98–102.
43 Riechson RB, Stander PE: Allergic bronchopulmonary aspergillosis: An increasingly common disorder among asthmatic patients. Post-grad Med 1988;88:217–219.
44 Slavin RG: ABPA in CF: A devasting combination. Pediatr Pulmonol 1996;21:1–2.
45 Patterson R: Potential errors in the diagnosis of ABPA; in Patterson R, Greenberger PA, Roberts M (eds). Allergic Bronchopulmonary Aspergillosis. Rhode Island, Oceanside, 1995, pp 29–33.
46 Greenberger PA: Allergic bronchopulmonary aspergillosis; in Middelton E, Reed CE, Ellis EF, Adkinson NF, Yunginger JW (eds): Allergy, Principle and Practice. St Louis, Mosby, 1988, pp 1219–1236.
47 American Academy of Allergy, Asthma and Immunology: Position statement: The use of standardized allergen extracts. J Allergy Clin Immunol 1997;99:583–586.
48 Norman PS: WHO-IUIS international standards: Advantage of these extracts; in Kurth R, Haustein D (eds): Regulatory Control and Standardization of Allergenic Extracts. 7th Int Paul Ehrlich Seminar. Stuttgart, Fischer, 1994, pp 59–64.
49 Nikolaizik WH, Crameri R, Blaser K, Schöni MH: Skin test reactivity to recombinant *Aspergillus fumigatus* allergen I/a in patients with cystic fibrosis. Int Arch Allergy Immunol 1996;111:403–408.
50 Moser M, Crameri R, Brust E, Suter M, Menz G: Diagnostic value of recombinant *Aspergillus fumigatus* allergen I/a for skin testing and serology. J Allergy Clin Immunol 1994;93:1–11.
51 Vijay HM, Perlmutter L, Bernstein IL: Possible role of IgG4 in discordant correlations between intracutaneous skin tests and RAST. Int Archs Allergy Appl Immunol 1978;56:517–522.
52 Dreborg S, Frew A (eds): Allergen standardization and skin tests. Allergy 1993;48:49–82.
53 Scheiner O, Kraft D: Basic and practical aspects of recombinant allergens. Allergy 1995;50:384–391.
54 King TP, Hoffman D, Lowenstein H, Marsh DG, Platts-Mills TAE, Thomas W: Allergen nomenclature. J Allergy Clin Immunol 1995;96:5–14.
55 Crameri R, Suter M: Display of biologically active proteins on the surface of filamentous phage: A cDNA cloning system for selection of functional gene products linked to the genetic information responsible for their production. Gene 1993;137:69–75.
56 Crameri R, Jaussi R, Menz G, Blaser K: Display of expression products of cDNA libraries on phage surfaces: A versatile screening system for selective isolation of genes by specific gene-product/ligand interaction. Eur J Biochem 1994;226:53–58.
57 Suter M, Foti M, Ackermann M, Crameri R: In Kay B, Winter KJ, McCafferty J (eds): A cDNA Cloning System Based on Filamentous Phage: Selection and Enrichment of Functional Gene Products by Protein/Ligand Interactions Made Possible by Linkage of Recognition and Replication Functions. San Diego, Academic Press, 1996, pp 187–205.

58 Crameri R, Hemmann S, Blaser K: pJuFo: A phagemid for display of cDNA libraries on phage surface suitable for selective isolation of clones expressing allergens; in Sehon A, KayGlass KT, Kraft D (eds): New Horizons in Allergy Immunotherapy. New York, Plenum Press, 1996, pp 103–110.

59 Crameri R: pJuFo: A phage surface display system for cloning genes based on protein-ligand interaction; in Schaefer B (ed): Gene Cloning and Analysis: Current Innovations. Wymondham, Horizon Press, 1997, pp 29–42.

60 Mierendorf RC, Percy C, Yang RA: Gene isolation by screening λgt11 libraries with antibodies. Meth Enzymol, 1987;152:458–469.

61 Barbas CF III, Lerner RA: Combinatorial immunoglobulin libraries on the surface of phage (Phabs): Rapid selection of antigen-specific Fabs. Comp Meth Enzymol 1991;2:119–124.

62 Achatz G, Oberkofler H, Lechnauer E, Simon B, Unger A, Kandler D, Ebner C, Prillinger H, Kraft D, Breitenbach M: Molecular cloning of major and minor allergens of *Alternaria alternata* and *Cladosporium herbarum*. Mol Immunol 1995;32:213–227.

63 Hemmann S, Blaser K, Crameri R: Allergens of *Aspergillus fumigatus* and *Candida boidinii* share IgE-binding epitopes. Am J Respir Crit Care Med 1997;156:1956–1962.

64 Moser M, Crameri R, Menz G, Schneider T, Dudler T, Virchow C, Gmahl M, Blaser K, Suter M: Cloning and expression of recombinant *Aspergillus fumigatus* allergen I/a (rAsp f I/a) with IgE binding and type I skin test activity. J Immunol 1992;149:454–460.

65 Arruda KL, Platts-Mills TAE, Fox JW, Chapman MD: *Aspergillus fumigatus* allergen I, a major IgE-binding protein is a member of the mitogillin family of cytotoxins. J Exp Med 1990;172: 1529–1532.

66 Crameri R, Faith A, Hemmann S, Jaussi R, Ismail C, Menz G, Blaser K: Humoral and cell-mediated autoimmunity in allergy to *Aspergillus fumigatus*. J Exp Med 1996;184:265–270.

67 Mayer C, Hemmann S, Faith A, Blaser K, Crameri R: Cloning, production, characterisation and IgE cross-reactivity of different manganese superoxide dismutases in individuals sensitised to *Aspergillus fumigatus*. Int Arch Allergy Immunol 1997;113:213–215.

68 Pan S, Cole GT: Molecular and biochemical characterisation of a *Coccidioides immitis*-specific antigen. Infect Immun 1995;63:3994–4002.

69 Galgian JN, Ampen NM: Coccidioidomycosis in human immunodeficiency virus-infected patients. J Infect Dis 1990;162:1165–1169.

70 Müller U, Dudler T, Schneider T, Crameri R, Fischer H, Skrbic D, Maibach R, Blaser K, Suter M: Type I skin reactivity to native and recombinant phospholipase A2 from honey bee venom is similar. J Allergy Clin Immunol 1995;96:395–402.

71 Müller U, Fricker M, Wymann D, Blaser K, Crameri R: Increased specificity of diagnostic tests with recombinant major bee venom allergen phospholipase A_2. Clin Exp Allergy 1997;27:915–920.

72 Disch R, Menz G, Blaser K, Crameri R: Different reactivity to recombinant *Aspergillus fumigatus* allergen I/a in patients with atopic dermatitis or allergic asthma sensitised to *Aspergillus fumigatus*. Int Arch Allergy Immunol 1995;108:89–94.

73 Crameri R, Lidholm J, Menz G, Grönlund H, Blaser K: Automated serology with recombinant allergens: A feasibility study; in Sehon A, HayGlass TH, Kratf D (eds): New Horizons in Allergy Immunotherapy. New York, Plenum Press, 1996, pp 111–116.

74 Crameri R, Lidholm J, Grönlund H, Stüber D, Blaser K, Menz G: Automated specific IgE assay with recombinant allergens: Evaluation of the recombinant *Aspergillus fumigatus* allergen I in the Phamacia CAP System. Clin Exp Allergy 1996;26:1411–1419.

75 Banerjee B, Kurup VP, Phadnis S, Geenberger PA, Fink JN: Molecular cloning and expression of a recombinant *Aspergillus fumigatus* protein Asp f II with significant immunoglobulin E reactivity in allergic bronchopulmonary aspergillosis. J Lab Clin Med 1996;127:253–262.

Dr. Reto Crameri, Swiss Institute of Allergy and Asthma Research,
Obere Strasse 22, CH–7270 Davos (Switzerland)
Tel. +41 81 413 70 83, Fax +41 81 413 16 07, E-Mail crameri@siaf.unizh.ch

Brakhage AA, Jahn B, Schmidt A (eds): *Aspergillus fumigatus.*
Contrib Microbiol. Basel, Karger, 1999, vol 2, pp 57–68

Host Defense Mechanism in *Aspergillus fumigatus* Infections

Markus Schneemann[a], *Andreas Schaffner*[b]

[a] Department of Microbiology and Immunology, Stanford University School of Medicine, Stanford, Calif., USA;
[b] Department of Medicine, University of Zurich Medical School, Zurich, Switzerland

Invasive aspergillosis is a life-threatening disease that develops only in well-defined subsets of severely immunocompromised patients. Epidemiological data point to prolonged leukopenia and therapy with high doses of corticosteroids as the most important risk factors for invasive aspergillosis [1, 2]. Other categories of immunodeficiencies such as deficient T-cell functions of AIDS, complement or antibody deficiencies are not by themselves sufficient to put patients at risk for this disease [3, 4]. This article will summarize the data that show that the strong innate resistance to aspergillosis resides in mononuclear and neutrophil phagocytes independent from instructions by T or B lymphocytes. In some nonmammalian species the production of antimicrobial peptides appears to be of particular importance.

Properties of the Pathogen Relevant to Infection and Immunity

Aspergillus fumigatus, the species most frequently isolated in invasive aspergillosis, belongs to a family of saprophytic filamentous fungi that are found ubiquitously in most environments. Aspergilli produce millions of asexual spores from each of their plentiful fruiting heads that can be considered to be omnipresent in the aerospora based on the ease with which they become

Table 1. Pathogenic properties of *Aspergilli*

Prolific asexual sporulation
Light hydrophobic spores permit efficient aerosolization
Growth on many organic substrates permits prosperity in most environments
Germination and hyphal growth at elevated temperatures above 37–39 °C allows for invasiveness in warm-blooded animals
Size of spores from many *Aspergilli* of 3–5 μm is perfect for their deposition in terminal airways
Spore wall protects metabolically inactive spores from toxic phagocytic products (reactive oxygen intermediates, defensins)
Angiotropism
? Catabolic enzymes such as elastase and other proteases
? Mycotoxins
? Complement inhibiting phospholipids

airborne. The spores, called conidia, have a size of a few microns and are able to germinate into hyphae even under very poor conditions, e.g. phosphate-buffered saline supplemented with *D*-glucose [5]. Hyphae of *A. fumigatus* can, however, call on a whole array of enzymes, proteins and toxins enabling it to thrive on many organisms' substrates, at a wide range of temperatures, even at 37–39 °C, as well as under competitive hostile conditions.

The main pathogenic virtues of *A. fumigatus* therefore are its omnipresence, its small size permitting penetration into the deep airways and its readiness to transform into tissue-invasive hyphae under host conditions (table 1).

As an opportunistic pathogen that only causes disease accidentally when host defenses are impaired, *Aspergilli* have not developed sophisticated survival strategies, as have professional pathogens such as *Listeria* or *Yersinia*. Therefore, it is not surprising that despite considerable efforts, no distinct virulence factors have unequivocally been defined for *A. fumigatus* so far [6].

The studies of null mutants in *A. fumigatus* excluded alkaline proteases with elastase activity, the ribotoxin restrictocin and the hyphal growth factor AfchsE as virulence factors [8, 9]. *A. fumigatus* produces catalase, an enzyme that inactivates reactive oxygen intermediates secreted by phagocytes in response to hyphae. Therefore, catalase was a prime candidate as a virulence factor, but the disruption of the catalase gene of *A. fumigatus* did not influence pathogenicity in a mouse model of aspergillosis [10].

Interactions with extracellular matrix components like fibronectin [11], laminin and fibrinogen [12] as a means to adhere to the host tissues were described for *A. fumigatus* conidia in vitro, but their significance in the in vivo pathogenesis of invasive aspergillosis remains unproven [6]. On the other hand, studies in *A. nidulans* defined the ability of the fungus to synthesize *p*-amino-

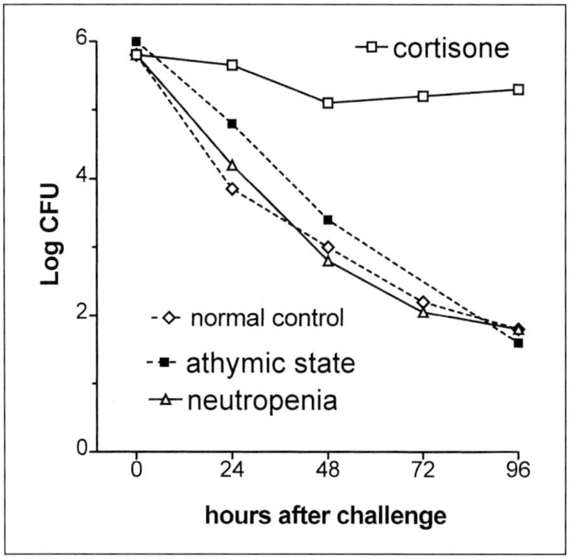

Fig. 1. Elimination of conidia from *A. fumigatus* by normal and immunodefficient mice. Groups of mice were challenged with aerosolized conidia from *A. fumigatus* and elimination of spores was followed by quantitative culture. Note that elimination follows a first-order kinetic without encouragement of killing by adoptive mechanisms. Adapted from Schaffner et al. [13].

benzoic acid as a crucial metabolic pathway that contributed to pathogenicity in a mouse model of invasive pulmonary aspergillosis [7].

The Host Response

Based on available epidemiological evidence it is host factors that determine whether *Aspergilli* establish disease. The encounter between the normal host and conidia is a daily event for everyone outside an environment protected by high-efficiency filters. People working in facilities managing organic waste inhale daily billions of spores including those from *A. fumigatus* but do not develop invasive aspergillosis. This is because the normal host efficiently eliminates conidia that reach terminal airways as aerosols by mechanisms of innate resistance residing in the lung.

In vivo studies in mice [13] have shown that resident pulmonary macrophages are responsible for clearing the lung from inhaled conidia of *A. fumigatus*, by a first-order kinetic (fig. 1). It is of note that conidia of *Aspergilli* have to become metabolically active and swell first before phagocytes are able to kill them [14]. This metabolic activation of conidia takes place when conidia encounter a suitable liquid environment, as in the terminal airways of the lung or inside the macrophage phagosome.

Human macrophages recognize and bind conidia by receptors that can be blocked by beta-1,-4-glucoside [15] or by a similar lectin-like interaction between conidia and an undefinced receptor of alveolar macrophages [16]. Phagocytosis is very efficient even in the absence of opsonic complement or immunoglobulins, an observation that is particularly important because these factors are not readily at hand in terminal airways. It is also for this reason that the in vivo significance for the in vitro demonstration of a more efficient phagocytosis in the presence of complement [17] seems to be of limited importance. A recent study, however, has proposed that pulmonary surfactant proteins enhance the rate of phagocytosis of *A. fumigatus* by human alveolar macrophages [18].

After phagocytosis, germination of conidia into hyphae is inhibited inside the phagolysosomal compartment, and blood-derived and alveolar macrophages from humans and various animal species kill and digest the spores [19]. After 24 h, 90% of engulfed conidia are killed, and 6–12 h later, completely digested [14].

The killing systems employed by the resident pulmonary macrophage to eliminate *A. fumigatus* conidia are not yet defined.

Reactive oxygen intermediates that readily kill bacteria like *Listeria* [20] do not appear to be involved in the killing process [21]. Human blood-derived macrophages that have been cultured for 10 days, and no longer produce significant amounts of reactive oxygen intermediates, are still able to kill conidia [22] as are macrophages from children with chronic granulomatous disease that do not dispose of a respiratory burst [5], an observation recently confirmed in knockout mice with chronic granulomatous disease devoid of a respiratory burst [23].

Nitric oxide and reactive nitrogen intermediates as products of the high-output inducible form of nitric oxide synthase, are not operative in the human and rabbit macrophage under conditions under which conidia are killed [22, 24], and in the mouse macrophage where nitric oxide synthase is functional, nitric oxide does not contribute to the killing of conidia [25]. Besides, high-output nitric oxide synthase activity is not only uninvolved in aspergillosis, but also in murine listeriosis [26]. Despite much controversy about the existence of the inducible nitric oxide synthase in the human phagocyte [27], to date no study has presented convincing data on high-output inducible nitric oxide synthase in human macrophages, i.e. metabolism of its substrate, turnover of its products or the availability of its obligate cofactor, tetrahydrobiopterin which is not synthesized in human macrophages [28].

Defensins, low molecular weight cationic peptide antibiotics, are found in many cell types including human neutrophils and rabbit alveolar macrophages [29], but are absent from human macrophages [5]. It is also of note that resident peritoneal macrophages from rabbits and mice do not possess efficient systems to control and kill ingested conidia from *A. fumigatus*, *A. niger* or *A. flavus* [22].

The Immunocompromise – Corticosteroids

Based on epidemiological data, pharmacological concentrations of corticosteroids are a major risk factor for the host to contract invasive aspergillosis [1, 2]. Accordingly, mice can be rendered highly vulnerable to aspergillosis by cortisone injections [13], and alveolar macrophages from humans and laboratory animals lose their capacity to control and kill conidia in vitro [22]. This suppressive effect is highly specific for corticosteroids, as cyclosporin A, a T-cell immunosuppressant, does not affect killing of *A. fumigatus* by macrophages [30].

Corticosteroids at pharmacological doses of 10^{-8} to 10^{-6} M impair killing of *A. fumigatus* conidia without interfering with phagocytosis, phagolysosomal fusion or the production and secretion of reactive oxygen intermediates [19]. The effect is mediated by the glucocorticoid receptor, as it is blocked by cortisone antagonists and starts only after 24–36 h of continuous treatment, indicating that the action of gucocorticoids is based on a catabolic hormone effect on the resident macrophage [31].

Neutrophils

The observation that prolonged neutropenia is a major risk factor for invasive aspergillosis is time honored [1, 2, 5]. Neutrophils, like macrophages, are inactive against conidia that have not been metabolically activated [13, 32]. The mechanisms seem to be (i) a low capacity of the cell wall of nonactivated conidia to trigger a respiratory burst and (ii) resistance to reactive oxygen intermediates [32]. It is of note that similar mechanisms seem to govern the susceptibility of spores and yeasts from virulent dimorphic fungi to the killing by neutrophils [33] improving their ability to survive in the host and to infect people with intact phagocytic defenses.

During germination and after transformation into hyphae *Aspergilli* become susceptible to killing by neutrophils. Neutrophils attack hyphae too large for phagocytosis, bind to their surfaces, release toxic oxygen radicals damaging

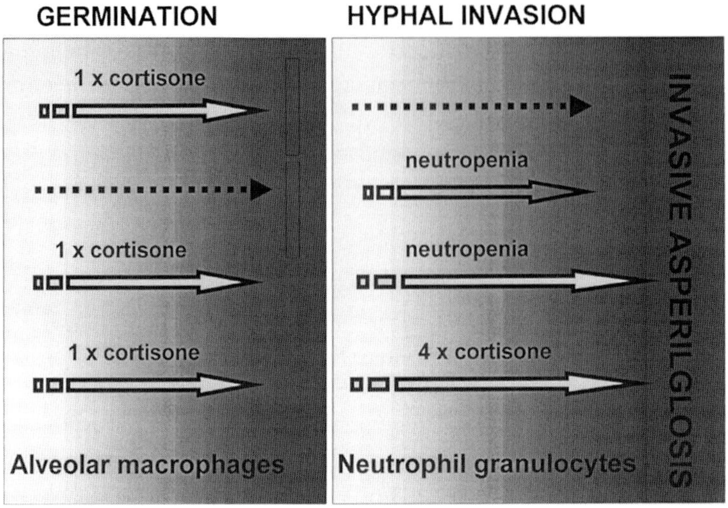

Fig. 2. Impairment of the high natural resistance to experimental aspergillosis of small laboratory animals (mice, rabbits or rats) by cortisone acetate or myelosuppressive drugs causing neutropenia. Note that the athymic state of 'nude' mice or cyclosporin A have no effect on anticonidial or antihyphal resistance (not shown).

[34] and killing the mycelial form of *Aspergilli* [13, 33]. The killing by neutrophils depends on reactive oxygen intermediates [33, 34] aided by the action of nonoxidative killing systems such as the neutrophil defensins (fig. 2). Besides prolonged leukopenia chronic granulomatous disease, a hereditary defect of oxygen radical formation of phagocytes is associated with a markedly increased risk for aspergillosis pointing to the importance of neutrophil-derived reactive oxygen intermediates for the control of *Aspergilli* [5].

Role for Cytokines in Aspergillosis

Interferon-gamma has been shown to activate macrophages to kill ingested bacteria like *Listeria* and *Salmonella* more effectively than nonactivated, resting phagocytes, mainly by enhancing their ability to produce reactive oxygen intermediates [21]. Macrophage-activating factors like interferon-γ and tumor necrosis factor-α, however, do not enhance the ability of human or mouse macrophage to eliminate *A. fumigatus* conidia [5, 21, 31]. The nonoxidative killing mechanisms operative against conidia appear to be unaffected by these cytokines.

Macrophage-deactivating factors like interleukin-4 and interleukin-10 have been demonstrated to suppress activated and resident macrophages in their capacity to kill bacteria [20]. Interleukin-10 deactivates the capacity of human blood-derived monocytes to kill *A. fumigatus* conidia, but this effect was smaller than that of corticosteroids [35]. It is likely that interleukin-10 diminishes not just the generation of reactive oxygen intermediates, but deactivates the same systems crucial for macrophages to control conidia. Interleukin-4, the most potent deactivator of macrophages in *Listeria* infections in vitro [20], had no effect on the killing of *A. fumigatus* [35]. Macrophage-colony-stimulating factor (M-CSF) has been shown to enhance phagocytosis of conidia and damage to hyphae by human monocytes and rabbit alveolar macrophages [36], but these effects could not be reproduced with monocytes elutriated from cancer patients treated with M-CSF [37].

Killing of *A. fumigatus* by human neutrophils could be enhanced in vitro by granulocyte-colony-stimulating factor with neutrophils from healthy volunteers [38], and by interferon-γ with neutrophils from patients with chronic granulomatous disease [39].

Taken together, activating cytokines have an augmenting effect on the killing of hyphae by both neutrophils and macrophages, but no activating signal has been shown to enhance killing of conidia by macrophages. Most importantly, no cytokine tested so far was able to counteract deactivation of macrophages against *Aspergilli* by corticosteroids [14, 35; Schaffner, unpubl. data].

The Antifungal Protein Drosomycin – A Lesson from Drosophila

Insects, despite not having an adaptive immune system are highly resistant to bacterial and fungal infections [40]. They depend on phagocytosis and capsule formation by blood cells and possess a large variety of peptides which are induced after challenge with either bacteria and fungi [40, 41].

When pricked with a needle containing activated *A. fumigatus* conidia, *Drosophila* secretes large amounts (10–100 µ*M*) of a single small peptide from their fat body, the insect's liver equivalent, into the hemolymph [41, 42]. This molecule, named drosomycin, turned out to be specific for fungi, and is not detectable in unchallenged or *Escherichia coli*-pricked insects [41, 42]. Studies with drosophila null mutants have identified the pathway by which this specific antifungal response is induced; it involves the spätzle/Toll/cactus pathway, the insect equivalent to the mammalian IL-1/NFkappaB cascade [41].

Drosomycin is a small, slightly cationic peptide highly resistant to proteases, heat and pH changes [42]. Its structure is similar to the insect defensins,

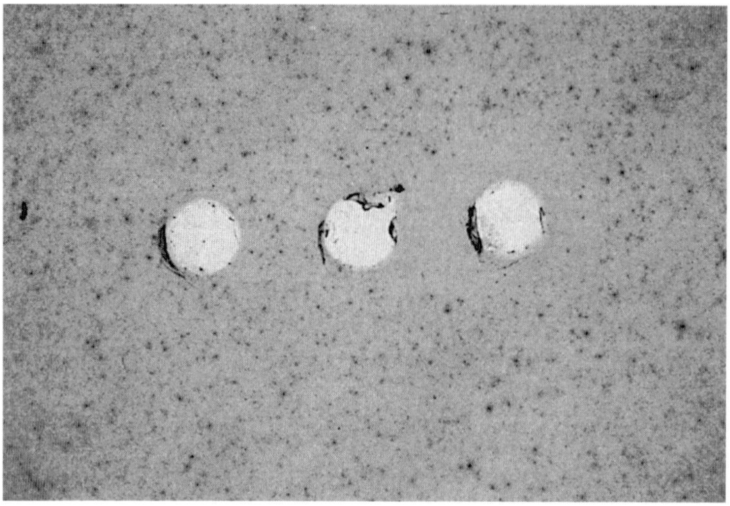

Fig. 3. Germinating conidia and hyphae from Aspergilli are highly susceptible to defensins, cationic natural antibiotics of human neutrophil granulocytes. Shown is the killing of germinated *A. fumigatus* by 100 ng/ml human defensins (HNP1-3) within 6 h of incubation at 37 °C added to the hole on the right side cut, into a poor plate of 2% agarose in phosphate-buffered glucose with 200 germinated spores/ml from *A. fumigatus*. Prior to the staining of the plate with methylene blue the plate was overlayered with trypticase soy agar and incubated for 12 h at 36 °C. Note the substantial *Aspergillus* growth inhibition zone around the well filled with defensins on the right, compared to the control well filled with phosphate-buffered saline only on the left.

with eight cysteine residues arranged in four intramolecular disulfide bridges, forming a central alpha helix linked to an antiparallel beta sheet [42, 43]. Drosomycin belongs to an ever-growing family of small cationic peptide antibiotics that are widespread among the plant and animal kingdom [40]. These peptide antibiotics kill by forming pores into the cell wall of the microbe, and some of them may even lead to a novel class of antibiotics for the treatment of human infectious diseases [44]. Studies of insect immunity and the apparent success of a nonadaptive, innate type of immunity already prompted some authorities in the field of mammalian adaptive immunology to call for a rethinking of immunology in general [45].

Apart from defensins found in human neutrophils [29], some new cationic peptides have recently been described in laboratory animals. Of interest in this context is the description of a human anti-leukoprotease as part of the mucosal defense in the respiratory tract against *A. fumigatus* conidia and *Candida albicans* [46]. It is produced by submucosal glands and epithelial cells in the bronchi and showed the same fungicidal potency as the human neutrophilic defensins HNP 1-3 (fig. 3) with a 50% conidia-killing efficacy of $5.6 \pm 0.7\ \mu M$ [46]. As with macrophages, and defensins, human anti-leukoprotease is not active against dormant, unmatured conidia. In order to be a significant host defense mechanism in aspergillosis, it should be postulated that its production or secretion is diminished by the main risk factor, corticosteroids. Its activity is dependent on low ionic strength, to such an extent that it will be inactive in the bronchial mucosal environment of patients with cystic fibrosis (CF).

CF, however, is not a risk factor for developing invasive aspergillosis [47], making it rather difficult to attribute a major role for antileukoprotease in the host defense against invasive *A. fumigatus*, but a lack of this protein might contribute to the ease with which *Aspergilli* colonize CF patients.

Conclusions

In its daily struggle against airborne *Aspergilli* the host can call upon two strong independent defense lines. The first is formed by alveolar macrophages that ingest, inhibit and kill the conidial form of the fungus independent from opsonins and without instructions by adaptive immune mechanisms. Killing of conidia does not call upon the production of reactive oxygen or nitrogen intermediates. The final outcome of the typical encounter between macrophage and conidium is its complete digestion. The second defense line is formed by neutrophil granulocytes (and quantitatively of minor importance by mononuclear phagocytes). Neutrophils attack hyphae that are too large for ingestion and kill them by oxidative and nonoxidative mechanisms such as defensins. Each defense line by itself is able to protect the host against a large conidial inoculum or over long time periods. Only if both defense lines are overcome, *Aspergillus* can cause invasive disease. Accordingly, prolonged neutropenia, defective neutrophil killing or pharmacological concentrations of glucocorticoids decrease anticonidial activity of macrophages and interfere with recruitment of neutrophils allowing the development of invasive aspergillosis. The mechanisms by which conidia are killed remain obscure, but neither nitric oxide nor oxygen radicals nor the defensins are involved. It is conceivable that human macrophages use small peptide antibiotics similar to defensins or drosomycin that are specifically induced upon contact with filamentous fungi.

References

1 Wald A, Leisenring W, van Burik JA, Bowden RA: Epidemiology of *Aspergillus* infections in a large cohort of patients undergoing bone marrow transplantation. J Infect Dis 1997;175:1459–1466.

2 Fridkin SK, Jarvis WR: Epidemiology of nosocomial fungal infections. Clin Microbiol Rev 1996; 9:499–511.

3 Schaffner A: Acquired immune deficiency syndrome: Is disseminated aspergillosis predictive of underlying cellular immune deficiency? J Infect Dis 1984;149:828–829.

4 Denning DW, Follansbee SE, Scolaro M, Norris S, Edelstein H, Stevens DA: Pulmonary aspergillosis in the acquired immunodeficiency syndrome. N Engl J Med 1991;324:654–662.

5 Schaffner A: Host defence in aspergillosis; in Bennett JE, Hay RJ, Peterson PK (eds): New Strategies In Fungal Disease. Edinburgh, Churchill-Livingstone, 1992, pp 98–112.

6 Bouchara JP, Tronchin G, Larcher G, Chabasse D: The search for virulence determinants in *Aspergillus fumigatus*. Trends Microbiol 1995;3:327–330.

7 Tang CM, Smith JM, Arst HN Jr, Holden DW: Virulence studies of *Aspergillus nidulans* mutants requiring lysine or p-aminobenzoic acid in invasive pulmonary aspergillosis. Infect Immun 1994; 62:5255–5260.

8 Smith JM, Tang CM, Van Noorden S, Holden DW: Virulence of *Aspergillus fumigatus* double mutants lacking restrictocin and an alkaline protease in a low-dose model of invasive pulmonary aspergillosis. Infect Immun 1994;62:5247–5254.

9 Aufauvre-Brown A, Mellado E, Gow NAR, Holden DW: *Aspergillus fumigatus chsE:* A gene related to *CHS3* of *Saccharomyces cerevisiae* and important for hyphal growth and conidiophore development but not pathogenicity. Fungal Genet Biol 1997;21:141–152.

10 Calera JA, Paris S, Monod M, Hamilton AJ, Debeaupuis JP, Diaquin M, Lopez-Medrano R, Leal F, Latge JP: Cloning and disruption of the antigenic catalase gene of *Aspergillus fumigatus*. Infect Immun 1997;65:4718–4724.

11 Penalver MC, O'Connor JE, Martinez JP, Gil ML: Binding of human fibronectin to *Aspergillus fumigatus* conidia. Infect Immun 1996;64:1146–1153.

12 Bouchara JP, Sanchez M, Chevailler A, Marot-Leblond A, Lissitzky JC, Tronchin G, Chabasse D: Sialic acid-dependent recognition of laminin and fibrinogen by *Aspergillus fumigatus* conidia. Infect Immun 1997;65:2717–2724.

13 Schaffner A, Douglas H, Braude A: Selective protection against conidia by mononuclear and against mycelia by polymorphonuclear phagocytes in resistance to *Aspergillus*: Observations on these two lines of defense in vivo and in vitro with human and mouse phagocytes. J Clin Invest 1982;69: 617–631.

14 Schaffner A: Macrophage-*Aspergillus* interactions. Immunol Ser 1994;60:545–552.

15 Kan VL, Bennett JE: Beta 1,4-oligoglucosides inhibit the binding of *Aspergillus fumigatus* conidia to human monocytes. J Infect Dis 1991;163:1154–1156.

16 Kan VL, Bennett JE: Lectin-like attachment sites on murine pulmonary alveolar macrophages bind *Aspergillus fumigatus* conidia. J Infect Dis 1988;158:407–414.

17 Kozel TR, Wilson MA, Farrell TP, Levitz SM: Activation of C3 and binding to *Aspergillus fumigatus* conidia and hyphae. Infect Immun 1989;57:3412–3417.

18 Madan T, Eggleton P, Kishore U, Strong P, Aggrawal SS, Sarma PU, Reid KB: Binding of pulmonary surfactant proteins A and D to *Aspergillus fumigatus* conidia enhances phagocytosis and killing by human neutrophils and alveolar macrophages. Infect Immun 1997;65:3171–3179.

19 Schaffner A, Schaffner T: Glucocorticoid-induced impairment of macrophage antimicrobial activity: Mechanisms and dependence on the state of activation. Rev Infect Dis 1987;9(suppl 5):S620–S629.

20 Bläuer F, Groscurth P, Schneemann M, Schoedon G, Schaffner A: Modulation of the antilisterial activity of human blood-derived macrophages by activating and deactivating cytokines. J Interferon Cytokine Res 1995;15:105–114.

21 Schaffner A, Rellstab P: Gamma-interferon restores listericidal activity and concurrently enhances release of reactive oxygen metabolites in dexamethasone-treated human monocytes. J Clin Invest 1988;82:913–919.

22 Schaffner A, Douglas H, Braude AI, Davis CE: Killing of *Aspergillus* spores depends on the anatomical source of the macrophage. Infect Immun 1983;42:1109–1115.
23 Morgenstern DE, Gifford MA, Li LL, Doerschuk CM, Dinauer MC: Absence of respiratory burst in X-linked chronic granulomatous disease mice leads to abnormalities in both host defense and inflammatory response to *Aspergillus fumigatus*. J Exp Med 1997;185:207–218.
24 Schneemann M, Schoedon G, Hofer S, Blau N, Guerrero L, Schaffner A: Nitric oxide synthase is not a constituent of the antimicrobial armature of human mononuclear phagocytes. J Infect Dis 1993;167:1358–1363.
25 Michaliszyn E, Senechal S, Martel P, de Repentigny L: Lack of involvement of nitric oxide in killing of *Aspergillus fumigatus* conidia by pulmonary alveolar macrophages. Infect Immun 1995;63:2075–2078.
26 Fehr T, Schoedon G, Odermatt B, Holtschke T, Schneemann M, Bachmann MF, Mak TW, Horak I, Zinkernagel RM: Crucial role of interferon consensus sequence binding protein, but neither of interferon regulatory factor 1 nor of nitric oxide synthesis for protection against murine listeriosis. J Exp Med 1997;185:921–931.
27 Albina JE: On the expression of nitric oxide synthase by human macrophages. Why no NO? J Leukoc Biol 1995;58:643–649.
28 Schneemann M, Schoedon G, Linscheid P, Walter R, Blau N, Schaffner A: Nitrite generation in interleukin-4-treated human macrophage cultures does not involve the nitric oxide synthase pathway. J Infect Dis 1997;175:130–135.
29 Ganz T, Lehrer RI: Antimicrobial peptides of leukocytes. Curr Opin Hematol 1997;4:53–58.
30 Berenguer J, Allende MC, Lee JW, Garrett K, Lyman C, Ali NM, Bacher J, Pizzo PA, Walsh TJ: Pathogenesis of pulmonary aspergillosis: Granulocytopenia versus cyclosporine and methylprednisolone-induced immunosuppression. Am J Respir Crit Care Med 1995;152:1079–1086.
31 Schaffner A: Therapeutic concentrations of glucocorticoids suppress the antimicrobial activity of human macrophages without impairing their responsiveness to gamma interferon. J Clin Invest 1985;76:1755–1764.
32 Levitz SM, Diamond RD: Mechanisms of resistance of *Aspergillus fumigatus* conidia to killing by neutrophils in vitro. J Infect Dis 1985;152:33–42.
33 Schaffner A, Davis CE, Schaffner T, Markert M, Douglas H, Braude AI: In vitro susceptibility of fungi to killing by neutrophil granulocytes discriminates between primary pathogenicity and opportunism. J Clin Invest 1986;78:511–524.
34 Diamond RD, Krzesicki R, Epstein B, Jao W: Damage to hyphal forms of fungi by human leukocytes in vitro: A possible host defense mechanism in aspergillosis and mucormycosis. Am J Pathol 1978;91:313–328.
35 Meier-Osusky I, Schoedon G, Bläuer F, Schneemann M, Schaffner A: Comparison of the antimicrobial activity of deactivated human macrophages challenged with *Aspergillus fumigatus* and *Listeria monocytogenes*. J Infect Dis 1996;174:651–654.
36 Roilides E, Sein T, Holmes A, Chanock S, Blake C, Pizzo PA, Walsh TJ: Effects of macrophage colony-stimulating factor on antifungal activity of mononuclear phagocytes against *Aspergillus fumigatus*. J Infect Dis 1995;172:1028–1034.
37 Roilides E, Lyman CA, Mertins SD, Cole DJ, Venzon D, Pizzo PA, Chanock SJ, Walsh TJ: Ex vivo effects of macrophage colony-stimulating factor on human monocyte activity against fungal and bacterial pathogens. Cytokine 1996;8:42–48.
38 Roilides E, Holmes A, Blake C, Venzon D, Pizzo PA, Walsh TJ: Antifungal activity of elutriated human monocytes against *Aspergillus fumigatus* hyphae: Enhancement by granulocyte-macrophage colony-stimulating factor and interferon-gamma. J Infect Dis 1994;170:894–899.
39 Ahlin A, Elinder G, Palmblad J: Dose-dependent enhancements by interferon-gamma on functional responses of neutrophils from chronic granulomatous disease patients. Blood 1997;89:3396–3401.
40 Boman HG: Peptide antibiotics and their role in innate immunity. Annu Rev Immunol 1995;13:61–92.
41 Lemaitre B, Nicolas E, Michaut L, Reichhart JM, Hoffmann JA: The dorsoventral regulatory gene cassette spätzle/Toll/cactus controls the potent antifungal response in *Drosophila* adults. Cell 1996;86:973–983.

42 Fehlbaum P, Bulet P, Michaut L, Lagueux M, Broekaert WF, Hetru C, Hoffmann JA: Insect immunity: Septic injury of *Drosophila* induces the synthesis of a potent antifungal peptide with sequence homology to plant antifungal peptides. J Biol Chem 1994;269:33159–33163.
43 Landon C, Sodano P, Hetru C, Hoffmann J, Ptak M: Solution structure of drosomycin, the first inducible antifungal protein from insects. Protein Sci 1997;6:1878–1884.
44 Hancock RE: Peptide antibiotics. Lancet 1997;349:418–422.
45 Medzhitov R, Janeway CA Jr: Innate immunity: The virtues of a nonclonal system of recognition. Cell 1997;91:295–298.
46 Tomee JF, Hiemstra PS, Heinzel-Wieland R, Kauffman HF: Antileukoprotease: An endogenous protein in the innate mucosal defense against fungi. J Infect Dis 1997;176:740–747.
47 Chung Y, Kraut JR, Stone AM, Valaitis J: Disseminated aspergillosis in a patient with cystic fibrosis and allergic bronchopulmonary aspergillosis. Pediatr Pulmonol 1994;17:131–134.

Prof. Dr. A. Schaffner, Medizinische Klinik B, Universitätsspital, CH–8091 Zürich (Switzerland)
Tel. +41 1 255 4020, Fax +41 1 255 4172, E-Mail klinsar@usz.unizh.ch

Brakhage AA, Jahn B, Schmidt A (eds): *Aspergillus fumigatus*.
Contrib Microbiol. Basel, Karger, 1999, vol 2, pp 69–87

Antigen and DNA Patterns Characteristic of *Aspergillus fumigatus*

J.P. Latgé

Aspergillus Laboratory, Pasteur Institute, Paris, France

Antigenic proteins and polysaccharides and DNA sequences can be used to characterize *Aspergillus fumigatus* [1]. Antigens play a major role in diagnosis since they are either passively released or actively produced in the host consecutively to fungal development. In both situations they promote a specific anti-*A. fumigatus* antibody response in the immunocompetent host or trace the presence of the fungus in the immunocompromised host where no antibodies are produced in response to fungal infection [2, 3]. DNA fingerprinting of *A. fumigatus* has been very useful for a good understanding of the ecology of this fungus as well as the epidemiology of its associated human diseases [4]. In addition, PCR assays have begun to show promising results for the diagnosis of *Aspergillus* infections [5, 6]. This chapter will review our current knowledge and understanding of protein, polysaccharide and DNA molecules useful for the identification and detection of *A. fumigatus*, inside and outside its human host.

Antigenic Molecules

Until the 1970s, *Aspergillus* infections were only seen in immunocompetent patients and resulted of limited mycelial colonization of the lung parenchyma. Serological diagnostic methods for such infections used simple immunochemical tests which main purpose was the detection of an increase of anti-*A. fumigatus* antibodies indicative of the reaction of the host against the infection. For such purposes, test or antigen were not required to be of a high quality. Antigenic preparations were often laboratory-produced and not standardized even when commercially available. Characterization of pure antigenic molecules was not attained at that time and the absence of antigen standardization has impeded an accurate quantification of the humoral immune response for a long time [1–3].

The situation has changed in recent years. Purification of antigens has become essential where one wishes to detect the presence of antigens in the biological fluids of patients with invasive aspergillosis (IA) or investigate difficult or underhand infections such as ABPA-like disease in cystic fibrosis patients. The selection of appropriate monoclonal antibodies (Mabs) to detect an antigen present in the biological fluids of patients with IA requires also the purification of the antigenic target. In a similar way, in order to identify antigens which play a role during the establishment of aspergillosis, these molecules first have to be isolated and characterized at a biochemical and genetical level. With the exception of their Ig binding properties, very little is known about the function of the different antigens during the infectious process. A restrictive definition of the word antigen would indicate that these molecules are produced by the fungus in the host during the development of the disease and stimulate an antibody response in a way that parallels the extension of the disease. This would also suggest that at least some of these antigens, particularly those with an enzymatic function, would help the fungus to penetrate and invade the host tissues.

Experimental conditions used to produce an antigenic extract influence greatly the pattern of antigens isolated from *A. fumigatus* [1–3, 7] (table 1). The variability of the antigenic preparation concerns not only the total protein and polysaccharide content but also the number and type of antigens present in each antigen batch. Amongst all the factors presented in table 1, the most critical seems the composition of the culture medium. The best antigenic preparations are obtained in media without sugar containing a single protein or a protein hydrolysate. The composition of such media seems closer to the nutritional environment encountered by the fungus in the lung, i.e. a protein-based medium (mainly composed of collagen and elastin) and a pH close to 7.4. The choice of the medium is also determinant when expression libraries are used to isolate clones which express antigens recognised by patient's sera [8, 9]. The antigen(s) identified will be highly dependent on the culture conditions used since the cDNA library constructed will reflect the mRNAs mostly expressed by the fungus under these culture conditions. In addition, comparative studies have suggested that the major antigens expressed in vitro may be different from the ones expressed in vivo during the colonization of host tissues [10, 11]. More work should be done in this area, since the quantification of antibodies directed specifically against antigens produced in the lung matrix, would increase the predictive values of the immunological tests.

Table 1. Source of variations in the composition of in vitro antigenic extracts [adapted from refs 1 and 7]

Important
Source
Conidium vs. mycelium
Intracellular vs. extracellular
Disrupted vs. undisrupted cells
Extraction buffer
pH
Detergent
Salt concentration
Composition of the culture medium
Defined vs. protein hydrolysate-based medium
Presence of glucose
pH
Duration of growth
Linear growth phase vs. stationary phase vs. sporulating phase (associated with mycelial lysis)
Not important
Growth temperature (room temperature vs. 37 °C)
Strain

In recent years, the extensive use of western blotting techniques performed after 1- or 2-dimensional electrophoresis has shown that about a hundred (glyco) proteins from *A. fumigatus* can bind human Ig [12–16]. Considered as a progress at the beginning, SDS-PAGE studies have brought more confusion than light to the understanding of the antigenic map of *A. fumigatus* for the following reasons: (i) the antigenic molecules are only identified on the basis of their M_r without in most cases, further purification to homogeneity and characterization of the biochemical function of the protein or sequencing of the encoding gene; (ii) authors usually do not take into account the putative partial degradation of the antigen during culture and extraction (by glycosidases or proteases) as well as a possible cross-reactivity of the antibodies with related but different antigenic molecules. As a result, molecules with the same M_r may correspond to 2 different antigens whereas other antigens with different M_r correspond to the same molecule processed differently by autolysis during the extraction procedure.

The definite identification of an antigen will only come from sequencing data (table 2). These data justify the efforts made in the last years to isolate and characterize at a molecular level polypeptide antigens responsible for a

Table 2. Antigenic proteins of *A. fumigatus* with known function

M_r (SDS-PAGE)[a]	Biochemical function	Gene cloned	Recombinant protein
18[b]	ribonuclease	+	+
19 (67)	superoxide dismutase	–	NA
27	superoxide dismutase	+	+
33	serine protease	+	+
38	aspartic protease	+	+
40	metalloprotease	+	+
82	metalloprotease	+	–
88[c]	dipeptidyl peptidase	+	+
88	heat shock protein	–	NA
90 (350)	catalase	+	+
94 (180)	dipeptidyl peptidase	+	+

NA = Not appropriate.
[a] Molecular mass of the native protein is indicated between brackets.
[b] Also known as Aspf1, Ag3.
[c] Also known as chymotrypsic antigen, Ag13, AgC.

specific antibody response. In addition, the availability of the encoding gene will allow for the production of recombinant antigens. A dozen of recombinant antigens of *A. fumigatus* have been listed in the literature and their antigenicity is similar to the native molecules. Two heterologous hosts have been successfully used, *Escherichia coli* and *Pichia pastoris* [9, 17–19]. The latter has the advantage to respect the glycosylation of the antigen which can be produced extracellularly. These recombinant molecules will allow in the near future a comprehensive study of the humoral immune response including the quantification of the different immunoglobulin isotypes in different pathological situations.

Protein Antigens with Known Enzymatic Activity

Only 3 types of enzymatic activity (RNAse, protease and oxidase) have been associated with antigenic molecules of *A. fumigatus* (table 2).

One of the major antigens of *A. fumigatus*, known also as Aspf I and Ag3 is an 18-kD ribonuclease binding to both IgE and IgG [20, 21]. This 18-kD antigen cleaves a single phosphodiester bond in a highly conserved region and releases a 300- to 400-base fragment from the 3′ end of the large ribosomal RNA [22]. By doing so, it interferes with the elongation factors and inactivate ribosome and protein synthesis in the cell at 10^{-9} *M*. The protein is composed of 149 amino acids with a 27-amino acid leader sequence and a putative active site composed of the 6 amino acids H_{49}, E_{95}, F_{96}, P_{98}, R_{120} and H_{136} [23]. On the basis of its enzymatic activity and amino acid sequence, this protein has been shown to be identical to restrictocin, a ribotoxin isolated in the early 1960s from *A. restrictus.* This molecule and α-sarcin, another related *Aspergillus* ribotoxin, have extremely powerful cytotoxic effect in the n*M* concentration range in an immunoconjugate form or on cells with altered permeability whereas killing of unmodified cells occurs at the μ*M* level [24–26]. The ribotoxin is present in the conidial cell wall and is secreted in vivo, as shown by its detection in the urine of patients with IA and in the regions of necrosis surrounding the fungus in the tissues infected with *A. fumigatus* [15, 27]. However, Aspf1 does not seem to play a major role in the pathogenicity of *Aspergillus* infections since mutants constructed by disruption of the gene encoding Aspf1 were indistinguishable from parental strains in their growth in lung tissues and mortality in a murine model of IA [28]. Understanding the exact nature of the effect of this toxin may be more complicated or subtle than originally thought. Preliminary experiments have shown a positive cumulative toxic effect when mixed lymphocyte populations are treated sequentially with subinhibitory doses of Aspf1 and gliotoxin, a toxic secondary metabolite produced by *A. fumigatus* [29, 30]. In addition, a mutated Aspf1 protein (obtained by the replacement of His at position 136 by Leu) which does not display RNAse activity [23] but is still antigenic, stimulates in a much higher way the immune response than the native toxic Aspf1. Such experiments show that the effect of a putative factor may look different when tested alone or in association with other *Aspergillus* molecules and that two functions displayed by the same molecule can interfere and stimulate different host responses [7].

The majority of the antigens characterized at a molecular level has a proteolytic activity. Indeed, these molecules were studied at the beginning as putative virulence factors [see chapter by Monod et al., this vol.]. The major protease secreted by *A. fumigatus* in presence of a protein or protein hydrolysate at neutral pH is the 33-kD serine protease ALP [31]. This protease of the subtilisin family which has been characterized biochemically and genetically does not seem to be a discriminative antigen for aspergillosis [32]. The study of the ALP mutant has allowed for the identification of a 40-kD metalloprotease (MEP) of the NPI family. Anti-MEP antibodies are produced in higher amount than anti-ALP antibodies in infected patients and seem discriminatory for aspergilloma sera [33]. These data suggested that MEP was either more antigenic or secreted in higher amount than ALP during lung invasion by *A. fumigatus.* Two other antigenic metalloproteases, also characterized by the HEXXH motif in which the two H residues coordinate the Zn^{2+} cation in the active centre and the E residue is involved in the bond breaking process, have been also isolated from *A. fumigatus* [34]. One is an 82-kD thimet oligopeptidase of the M3 family similar to the yeast saccharolysine and the mammalian thimet oligopeptidase [35]. It is recognized by human antibodies from aspergilloma patients but its diagnostic potential has not been investigated (Ibrahim-Granet, unpublished). The other metalloprotease is a 20-kD protein of the NPII family which antigenicity has not been investigated [36]. A third protease (PEP) produced by *A. fumigatus* has been characterized recently. It is an aspartic proteinase of 38 kD of the aspergillopepsin family, D_{32} and D_{214} being recognized as the active site [37]. Anti-PEP antibodies are present in sera of patients suffering from aspergillosis and PEP has been demonstrated by immunofluorescence in human lung infected by *A. fumigatus*. In contrast to ALP and MEP, the expression of PEP does not seem to be induced by collagen or elastin. It is worth noting that another *A. fumigatus* protein tightly bound to the cell wall cross-react with anti-PEP antibodies [38].

Amongst all the proteases, the ones which look the most antigenic belong to the dipeptidylpeptidase (DPP) family with the G-X-S-X-G consensus sequence of the serine proteases. One of these proteases is a 180-kD dimeric DPP IV with a 94-kD subunit [39; Debeaupuis, unpubl.]. This DPP which releases X-Pro dipeptides from the NH_2 terminal end of the substrate, is homologous to the CD26 and has a putative catalytic site S_{613}, D_{690}, H_{725}. The reactivity of this protein towards aspergilloma patients has been noted but not quantitatively estimated. The other dipeptidylpeptidase isolated releases mainly X-Ala dipeptides (but also His-Ser and Ser-Tyr) from the N-terminus of polypeptides [17]. The gene of this protein encodes for a protein of 79 kD with a signal peptide of 18 amino acids. This result is in agreement with biochemical data showing also that this dipeptidylpeptidase migrates as a protein doublet of 87–88 kD and contains an approximately 9-kD N-linked carbohydrate moiety. The biochemical properties of this enzyme as well as its exocellular localization indicates it is a new enzyme which belongs to a new class of dipeptidylpeptidase (DPP V) with the putative catalytic triad of the DPP arranged as S_{560}, D_{643}, T_{675}. This protein which does not have any chymotrypsic activity is indeed the so-called chymotrypsic antigen which has been used for a long time in the serodiagnosis of aspergillosis in the immunocompetent host and has been defined in the past as a chymotrypsin only on the basis of a colorimetric reaction resulting from the release of naphtol radicals by a precipitin band in presence of the chromogenic substrate N-acetyl-phenylalanine naphtyl ester [17]. Its potential in the serodiagnosis of aspergilloma has been confirmed [40].

The third category of antigens with known enzymatic activity corresponds to oxidases. Two catalases and two superoxide dismutases have been described in *A. fumigatus*. A tetrameric 67-kD Cu, Zn superoxide dismutase has been shown to be useful for the serological diagnosis of aspergillosis infection in the immunocompetent host [41, 42]. This molecule has been biochemically purified but its encoding gene has not been cloned yet. The second superoxide dismutase has been identified through the screening of an *A. fumigatus* library displayed on the surface of filamentous phages. It is a 27-kD Mn-superoxide dismutase which presents 67% similarity to the corresponding human enzyme [8]. Interestingly, recombinant proteins from both the human and *A. fumigatus* enzymes are recognized by IgE antibodies from patients allergic to *A. fumigatus*, elicit immediate skin reactivity and promote proliferation of peripheral blood mononuclear cells. These observations provide evidence for autoreactivity to the human Mn SOD in allergic patients sensitized to an environmental allergen from *A. fumigatus* with a high degree of sequence homology to the corresponding human enzyme. Similar studies should be expanded to other human and *A. fumigatus* molecules displaying significant homologies such as CD26 and *A. fumigatus* dipeptidylpeptidase IV (DPP IV) or *A. fumigatus* metalloprotease B (MEPB) and thimet oligopeptidase [35, 39].

Catalase antibodies have been detected in serum samples of patients with aspergilloma. Proteins bearing this enzymatic activity have been indeed one of the first detector antigens used for the diagnosis of aspergilloma. Two catalase bands can be identified by electrophoresis on nondenaturing polyacrylamide gels [43]: a slow migrating band (S) which binds to Con A and a fast migrating band (F) which does not bind to Con A and in contrast to S, has an additional peroxidase function. Both molecules are antigenic with the S component being the most antigenic molecule. The S catalase of *A. fumigatus* is a tetrameric protein with a monomeric subunit of 90 kD [18, 44]. The oligomeric subunit contains an N-linked sugar moiety of 7 kD which does not bear the antigenic epitope of the molecule. It is an extremely resistant protein whose activity is not affected by high temperatures, reducing and denaturing agents. Analysis of the structural gene of the S catalase, *CAT1*, and its deduced amino acid sequence shows that *CAT1* has both a signal peptide of 15 amino acids and a propeptide of 12 amino acids with the pair of basic amino acids Arg_{26}-Arg_{27} acting as a cleavage site for a KEX2-like endopeptidase [18]. Comparison of *CAT1* sequence with other catalase genes suggests that the three amino acids involved in the binding to the heme prosthetic group are conserved in *CAT1* in the form of H_{120}, S_{141}, N_{175}. Immunoassays have shown that this protein can be used as a detector antigen to diagnose pulmonary aspergillosis in the immunocompetent host [44, 45].

Other Proteins

Other *A. fumigatus* antigenic proteins have been insufficiently characterized even though their potential in diagnosis seems promising, at least for some of them. Amongst them, I would mention: (1) A 28-kD exocellular protein of unknown function [Debeaupuis, unpubl.]. This protein is very immunogenic as shown accidentally when present as a contaminant of a batch of purified Aspf1. Its identity with the 29-kD antigen found in the urine of patients with IA [46] remains to be investigated. (2) A 20-kD extracellular [47] and a 93-kD intracellular protein [48]. (3) The unknown Ag5 of Longbottom which has not been purified yet but is an extremely fast released antigen [49]. The concept of timing in the release of antigens is very interesting since it may be useful in the diagnosis of IA where the detection of the first antigens released may be synonymous of early detection. (4) An 88-kD heat-shock protein [10]. (5) The Ag7 complex. The recent molecular characterization of a major antigen of *A. nidulans*, ASPND1 [50], the use of monospecific antisera and a careful review of the literature seem to indicate that the antigens from *A. fumigatus* called p60, p40, p37 [16, 50], AspfII [9], gp55 [51], 41 and 53 kD of CS2 [52], 58 kD [53], 35–65 kD [54] and 66 kD [55] belong to the Ag7 family. It is not known whether these proteins are indeed the same molecule with different levels of phosphorylation and/or glycosylation, or represent a family of very related proteins sharing common epitopes. A characterization of the members of this complex is presently undertaken.

Polysaccharidic Antigens

Galactomannan (GM) is an essential component of the cell wall of *A. fumigatus* and other *Aspergillus* species. It is also actively secreted during fungal growth in vitro where it can reach a concentration of 200 μg/ml of culture medium [56]. The antigenicity of this molecule has been demonstrated 30 years ago by several Japanese studies [57]. GM of *A. fumigatus* with different origins has been chemically analyzed. Although data slightly differ, a consensus structure can be established: the mannan core has a linear configuration containing α(1-2)- and (1-6)-linked residues in the ratio of 3:1 and the antigenic acid-labile side chains branched on two α(1-2)-linked mannose residues are composed of β(1-5) galactofuranosyl residues with an average degree of polymerisation of 4 [56, 58, 59]. Numerous intra- and exocellular glycoproteins of *A. fumigatus* with $M_r \geq 40$ kD bear also this galactofuran epitope. The type of glycosylation involved in the linkage of GM to proteins has not been studied.

The galactomannan has been historically the first antigen detected in experimentally infected animals and in patients with IA [60]. In spite of some contradictory results which may be linked to the level or the regimen of immunosuppression in relation to the underlying disease of these patients, most previous studies indicate that *A. fumigatus* infections in an immunosuppressed host is not accompanied by an increase in anti-*A. fumigatus* antibody titers [61–64]. The detection of the antigens in the body fluid of the patient (sera, urine and bronchoalveolar lavage (BAL)) remains the only laboratory diagnosis method useful for this disease [1]. The major difficulty associated to the detection of antigen is the very low amount of antigen present in the biological fluids (ng or <ng quantities per ml) [65]. Until now, even though a few antigenic proteins were identified in the sera or urine of these patients [46], the only antigen which retains its antigenicity through the treatment for dissociating the immune complexes and is present in sufficient amount to be detectable in a way useful for the clinic is GM. The selection of anti-GM Mab recognizing the β1-5 galactofuran side chain of GM has been at the basis of the development of the unique commercial test available to diagnose IA serologically [65]. It is a sandwich ELISA able to detect 0.5–1 ng/ml of GM in a biological sample [66]. Even though the sensitivity of this method remains insufficient, this test has shown its efficacy in the clinical setting [67, 68] and has demonstrated repetitively the occurrence of IA in the absence of clinical signs [see chapter by Kappe, this vol.].

β1-3 Glucan, which is another component of the fungal cell wall, can also be used as a detector antigen for IA even though it is not an immunogenic molecule. In this case, the detection system is based on the activation of a proteolytic coagulation cascade which is an essential defense mechanism of arthropods against the invasion of pro- and eucaryotic microorganisms in their hemocecle [69].

DNA Sequences

Diagnosis

DNA of *A. fumigatus* can be detected by PCR in patients with IA. Initial studies were centered on BAL and clinically confirmed IA was always associated retrospectively with a positive PCR test [70, 71]. The different DNA probes used successfully originated from single copy genes such as the 18-kD ribotoxin or the 33-kD alkaline protease or from repeated sequences (18S or 26S rDNA or mitochondrial DNA) [1]. Due to the continuous occurrence of conidia in the atmosphere, false-positive results were generated by the presence in the respiratory tract of airborne conidia which are not causing active disease: up to 25% of positive detections have been obtained from BAL samples of control patients [72, 73]. The number of false-positive samples was higher using PCR assays in BAL than ELISA for antigen detection [74]. Recently, very promising PCR results were obtained using blood (serum or plasma) samples [5, 6]. This line of investigation should be pursued since PCR detection in blood has several obvious advantages over BAL: (i) no false-positives are obtained due to the absence of external contamination; (ii) blood sampling is easier to perform than BAL which is always technically or ethically difficult in patients at high risk for IA; (iii) samplings can be repeated and will, in the near future, allow a quantification of the PCR detection in a similar way as with ELISA. When compared to ELISA, PCR positivity in blood seems to occur always later than GM detection [Bretagne, pers. commun.]. However, the combined use of PCR and ELISA should allow a definitive diagnosis of IA even in absence of obvious clinical symptoms. In addition, a comparative evaluation of PCR and ELISA data should allow for a better understanding of transient aspergillosis which seems to occur in neutropenic patients in absence of clinical symptoms and which would be responsible for ELISA data considered up to data as false-positive. The PCR results obtained in blood samples raise also an interesting question on the origin of the DNA amplified since *A. fumigatus* is not cultured from blood, even in the late stages of the disease.

Strain Fingerprinting in A. fumigatus

The rationale for typing isolates of *A. fumigatus* has a clinical relevance. Obviously, the infectious inoculum for aspergillosis are airborne conidia which are ubiquitous and present at an average concentration varying from 0.1 to 10 conidia/m^3 of air. Indeed, outbreaks of IA have been often associated in the past with a transient increase in the concentration of airborne spores [1]. To reduce the level of air contamination and the risk of infection, air filtration devices have been constructed in units at risk for IA [75]. However, reaching a level zero of contamination is utopic and can be only attained in very limited areas (e.g. laminar flow bed). Indeed, many IA cases still occur under laminar flow. In addition, such a system of protection has not solved the problem of late infections [76]. The blind application of air filtration prophylactic measures remains insufficient for a population at risk for IA. It was essential to identify or at least understand the ecological reservoir of the infectious inoculum in view to target most efficiently the measures of protection for the patient. Such study requires a very precise method of strain fingerprinting. This method must be both highly discriminative, reproducible and independent of the environment where the fungus grows. For these reasons, phenotypic methods based on protein patterns detected by antibodies or enzymatic [77, 78] substrates should be discarded because they do not fulfill the criteria mentioned above: proteins produced are highly depending on the culture conditions and protein patterns can be used at the most to rank strains at a subspecies level. In contrast, genotypic methods are highly independent of the external milieu. Three methods to fingerprint the DNA of *A. fumigatus* strains have been developed up to date: two use the polymerase chain reaction (PCR) with different types of primers and principles, whereas one is based on restriction fragment length polymorphisms (RFLP) vizualized after hybridization with a specific probe.

Typing Methods. Hybridization of restriction enzyme DNA fragments with repeated DNA sequences, a method successfully used for typing of other fungal pathogens, has been recently introduced to type *A. fumigatus* strains and has given to date unique strain differentiation [4, 79]. This RFLP method uses a species-specific repeated sequence λ 3.9 as a marker and provide unique *Eco*RI and *Hin*dIII Southern blot hybridization patterns with >25 bands for each strain tested. Hybridization patterns with the repeated sequence are amenable to computer-aided analysis which makes possible the comparative analysis of thousands of strains. The repeated sequence *Afut*1 isolated from the repetitive DNA sequence λ 3.9 used for strain fingerprinting is an inactive retroelement of 6.9 kb bounded by two long terminal repeats (LTR) of 282 bp, with sequences and features characteristic of retroviruses and retrotransposons [80]. Comparison of the peptidic sequences with other LTR retrotransposons showed that *Afut*1 is a member of the gipsy family identified originally in *Drosophila*. Multiple copies of the same or homologous *Afut*1 element are found in the genome of *A. fumigatus*.

Random amplification of polymorphic DNA (RAPD) by PCR relies on primers of arbitrary sequence to amplify segments of genomic DNA in a polymerase chain reaction. Although many primers have been evaluated only a small proportion of those examined could detect variability among *A. fumigatus* strains [81–83]. Amongst them, the primer R108 (GTATTGCCCT) identified by Aufauvre-Brown et al. [82] has shown the best discriminatory power. One of the problems associated with RAPD and inherent to the technique due to the low annealing temperature using arbitrary sequences is the lack of reproducibility of the patterns obtained. The source of *Taq* DNA polymerase and the magnesium ion concentrations have been shown to be key variables in producing a good reproducible discrimination. If the *Taq* DNA polymerase is not carefully titrated against the magnesium ion concentration, loss of discriminatory power may occur potentially generating 'pseudoclusters' or variations in band intensity [81, 84].

Microsatellites are stretches of short-repetitive sequences distributed across eukaryotic genomes. Because of the great variability of repeat number at most loci, microsatellites have been used widely in genetic mapping and as population markers. If thousands of microsatellite loci have been cloned from the genome of higher plants and animals, microsatellites in the genome of pathogenic fungi just begin to be studied [85]. In absence of available microsatellite sequences in data banks, the first attempt to isolate microsatellites in

A. fumigatus consisted of screening libraries of digested total DNA from *A. fumigatus* with a $(CA)_{10}$ oligonucleotide. Four clones contained long CA repeats $[(CA)_9(GA)_{25}]$, $[(CA)_2C(GA)_{23}]$, $[(CA)_8]$ and $[(CA)_{21}]$. Four sets of primers were designed to amplify these microsatellites. One primer of each set was 5′ labeled with fluorescein to allow sizing of PCR products with an automatic sequencer. Eight, 10, 10 and 17 alleles were found for the four microsatellites, respectively, in studying 50 independent isolates and the combination of the four microsatellite markers allowed for the identification of 38 different associations amongst these isolates [86]. Each typing method has his pros and cons supporters. A comparative trial is presently undertaken with these 3 methods to select the most discriminative one or see if their association provides a better efficacy in the identification of strains of *A. fumigatus* than a single method as it has been already shown in *A. fumigatus* and other fungal pathogens [87, 88].

Molecular Epidemiology of Invasive Aspergillosis. RAPD typing has been used in the past to demonstrate or exclude a nosocomial acquisition of*Aspergillus* infection in patients with IA. However, these studies were always limited to a very small number of isolates usually obtained during a cluster of infection [89, 90].

The most extensive and advanced studies in the analysis of *A. fumigatus* have been performed using the repeated DNA sequence λ 3.9 mentioned above [91, 92]. Using this probe, a fingerprinting analysis of about 2,000 isolates of *A. fumigatus* with different origins from all over the world has shown the extremely high genetic diversity of this fungal species [4]. The high genetic diversity of an *A. fumigatus* population which is seen when strains from all over the world are compared is also found in a restricted environment such as a single hospital. In the hospital Hotel Dieu (Paris, France) where the *A. fumigatus* population was followed for a period of 2 years, around 85% of the strains were recovered only once. Only 10–15% of the genotypes were found at least twice and some of these strains can persist for prolonged periods in different buildings of the same hospital [91]. It was never found a strain population specific of a particular location which would have resulted from the development of the fungus in this specific location. When 400 strains from 2 hospitals of Paris distant of 3 miles were compared pairwise, only 8% of identities were found between strains of the two hospitals while a value of 3% was seen when these strains were compared to a non-Paris population. This result suggests, that in spite of its extreme heterogeneity, the source of the *fumigatus* population originates from the own environment of the hospital itself with a limited contamination from the overall Paris atmosphere.

Analysis of hundreds of clinical and environmental isolates of *A. fumigatus* using this probe demonstrated that no genetical discrimination can be made on the basis of the saprophytic or clinical origin of the isolates [4]. This result means that every isolate present in the environment can become pathogenic if it encounters the appropriate host. It has some practical implication since it will indicate that the prevention measures should be applied to any environmental *A. fumigatus* conidia.

In the hospital setting, the main purpose of fingerprinting *A. fumigatus* isolates is to investigate the nosocomial acquisition of *Aspergillus* infections. The identity found in numerous occasions between patient and environment isolates suggests that most patients have acquired their infection in hospital [91]. However, even during an aspergillosis outbreak, it is very rare to encounter several patients infected by the same strain. This result is in agreement with the typing data of the environmental studies indicating that each patient is surrounded by an extremely diverse population of strains. So, the chances for two patients to be infected by the same strain are very low. If IA occurs 3 months after bone marrow transplantation, it can be calculated from our data that every patient has inhaled during this period at least 6,000 different genotypes. Accordingly, the absence of identity between genotypes found in patients and the environmental strains does not exclude the nosocomiality of an infection. It only indicates that the environmental population typed only reflects a limited portion of the fungal population actually breathed by every patient. Assessing a nosocomial infection remains difficult and requires the molecular typing of several isolates per patient as well as hundreds of strains from the environment collected from a prolonged period of time to be able to find, due to the huge diversity of the outside population, the contaminating environmental strain. Nosocomiality of IA can still be demonstrated after several months of incubation of the fungus in the patient or during the relapse of the underlying disease. Such eventuality must be carefully checked before any IA infection could be assigned to a community-acquired case.

Conclusion

Significant progress has been made in recent years in the genotypic and phenotypic characterization of *A. fumigatus*. Strain fingerprinting has allowed a good understanding of the genetic diversity existing in the species *fumigatus*. It also suggested that every strain has the same infectivity potential. The absence of environmental strains of *A. fumigatus* with reduced aggressiveness is in agreement with previous biochemical, molecular as well as immunological studies which have been unable to identify a key factor responsible for the pathogenicity of this fungal species [1]. These results question the occurrence of prominent or single putative virulence factors in a saprophytic species such as *A. fumigatus* which is not submitted to any selective pressure in humans since it does not need this host to complete its biological cycle.

However, during infection the fungus releases in its human host antigenic molecules which may play a role in the host-fungus interactions. Until now these molecules have been only investigated because of their usefulness in the diagnosis of *Aspergillus* infections. Although anti-*A. fumigatus* antibodies can still be demonstrated using crude extracts and simple immunoelectrophoretic methods, the recent purification of the 3 major *A. fumigatus* antigens (Aspf1, catalase and dipeptidyl peptidase V) and their availability as recombinant antigens, will allow now for a quantitative study of the humoral response of the immunocompetent host against this fungus. Diagnosis of invasive aspergillosis remains more problematic in the immunocompromised host. A positive detection of galactomannan associated to clinical signs is synonymous of infection. In absence of clinical symptoms, a positive antigen detection associated to a positive PCR assay can be now considered as a positive diagnosis. However, efforts should be directed now in view to increase (a) the sensitivity of the methods, for example through the use of the immuno-PCR technique, and (b) the specificity of the detection consecutively to the identification of other antigens than GM released in the early phases of the infection.

There is no clue today of the reasons for an *A. fumigatus* protein or polysaccharide to become antigenic. Some of the discriminatory antigens are produced by *A. fumigatus* in high amount but this is not a general rule. Up to date, it is still unknown if *A. fumigatus* antigens play a role during infection or if these molecules are just a flag signaling the presence of the fungus in the host. No studies have indicated a role for specific antigens in the stimulation of T cells which have been shown recently to be essential in the healing/non-healing response to an *A. fumigatus* infection [93]. Similarly, the putative role of antigens in protection against the infection has not been thoroughly studied. The induction of a specific isotype response linked to different *Aspergillus* pathologies as well as the role of different epitopes in inducing a different Th1/Th2 response are just emerging topics [1]. All the questions mentioned above can now be approached through the use of the different recombinant antigens available today.

References

1 Latgé JP: *Aspergillus fumigatus* and aspergillosis. Clin Microbiol Rev 1999, in press.

2 Hearn VM: Antigenicity of *Aspergillus* species. J Med Vet Mycol 1992;30:11–25.

3 Kurup VP, Kumar A: Immunodiagnosis of aspergillosis. Clin Microbiol Rev 1991;4:439–456.

4 Debeaupuis JP, Sarfati J, Chazalet V, Latgé JP: Genetic diversity among clinical and environmental isolates of *Aspergillus fumigatus*. Infect Immun 1997;65:3080–3085.

5 Einsele H, Hebart H, Roller G, Löffler J, Rothenhöfer I, Müller CA, Bowden RA, Van Burik J-A, Engelhard D, Kanz L, Schumacher U: Detection and identification of fungal pathogens in blood by molecular probes. J Clin Microbiol 1997;35:1353–1360.

6 Yamakami Y, Hashimoto A, Tokimatsu I, Nasu M: Detection of DNA specific for *Aspergillus* species in serum of patients with invasive aspergillosis. J Clin Microbiol 1996;34:2464–2468.

7 Latgé JP, Paris S, Sarfati J, Debeaupuis JP, Monod M: Exoantigens of *Aspergillus fumigatus:* Serodiagnosis and virulence; in Powell KA, Renwick A, Peberdy JF (eds): The Genus *Aspergillus:* From Taxonomy and Genetics to Industrial Application. London, Plenum Press, 1994, pp 321–340.

8 Crameri R, Hemmann S, Jaussi R, Ismail C, Menz G, Blaser K: Humoral and cell-mediated autoimmunity in allergy to *Aspergillus fumigatus.* J Exp Med 1996;184:265–270.

9 Banerjee B, Kurup VP, Phadnis S, Greenberger PA, Fink JN: Molecular cloning and expression of a recombinant *Aspergillus fumigatus* protein *Aspf* II with significant immunoglobulin E reactivity in allergic bronchopulmonary aspergillosis. J Lab Clin Med 1996;127:253–262.

10 Burnie JP, Matthews RC: Heat-shock protein 88 and *Aspergillus* infection. J Clin Microbiol 1991; 29:2099–2106.

11 Sarfati J, Boucias DG, Latgé JP: Antigens of *Aspergillus fumigatus* produced in vivo. J Med Vet Mycol 1995;33:9–14.

12 Baur X, Weiss W, Jarosch B, Menz G, Schoch C, Schmitz-Schumann M, Virchow C: Immunoprint pattern in patients with allergic bronchopulmonary aspergillosis in different stages. J Allergy Clin Immunol 1989;83:839–844.

13 Bernstein JA, Zeiss CR, Greenberger PA, Patterson R, Marhoul JF, Smith LL: Immunoblot analysis of sera from patients with allergic bronchopulmonary aspergillosis: Correlation with disease activity. J Allergy Clin Immunol 1990;86:532–539.

14 Hearn VM, Wilson EV, Proctor AG, Mackenzie DWR: Preparation of *Aspergillus fumigatus* antigens and their analysis by two-dimensional immunoelectrophoresis. J Med Microbiol 1980;13: 451–458.

15 Latgé JP, Moutaouakil M, Debeaupuis JP, Bouchara JP, Haynes K, Prévost MC: The 18-kilodalton antigen secreted by *Aspergillus fumigatus.* Infect Immun 1991;59:2586–2594.

16 Lopez-Medrano R, Ovejero MC, Calera JA, Puente P, Leal F: *Aspergillus fumigatus* antigens. Microbiology 1995;141:2699–2704.

17 Beauvais A, Monod M, Debeaupuis JP, Diaquin M, Kobayashi H, Latgé JP: Biochemical and antigenic characterization of a new dipeptidyl-peptidase isolated from *Aspergillus fumigatus.* J Biol Chem 1997;272(10):6238–6244.

18 Calera JA, Paris S, Monod M, Hamilton AJ, Debeaupuis JP, Diaquin M, Lopez-Medrano R, Leal F, Latgé JP: Cloning and disruption of the antigenic catalase gene of *Aspergillus fumigatus.* Infect Immun 1997;65:4718–4724.

19 Moser M, Menz G, Blaser K, Crameri R: Recombinant expression and antigenic properties of a 32-kilodalton extracellular alkaline protease, representing a possible virulence factor from *Aspergillus fumigatus.* Infect Immun 1994;62:936–942.

20 Arruda LK, Mann BJ, Chapman MD: Selective expression of a major allergen and cytotoxin, Asp f1, in *Aspergillus fumigatus.* Implications for the immunopathogenesis of *Aspergillus*-related diseases. J Immunol 1992;149:3354–3359.

21 Longbottom JL: Antigens and allergens of *Aspergillus fumigatus.* II. Their further identification and partial characterization of a major allergen (Ag 3). J Allergy Clin Immunol 1986;78: 18–24.

22 Lamy B, Davies J: Isolation and nucleotide sequence of the *Aspergillus restrictus* gene coding for the ribonucleic toxin restrictocin and its expression in *A. nidulans:* The leader sequence protects producing strains from suicide. Nucl Acids Res 1991;19:1001–1006.

23 Yang R, Kenealy WR: Effects of amino-terminal extensions and specific mutations on the activity of restrictocin. J Biol Chem 1992;267:16801–16805.

24 Madan T, Arora N, Sarma PU: Identification and evaluation of a major cytotoxin of *A. fumigatus.* Mol Cell Biochem 1997;167:89–97.

25 Munoz A, Castrillo JL, Carrasco L: Modification of membrane permeability during Semliki Forest Virus infection. Virology 1985;146:203–212.

26 Wawrzynczak EJ, Henry RV, Cumber AJ, Parnell GD, Derbyshire EJ, Ulbrich N: Biochemical, cytotoxic and pharmacokinetic properties of an immunotoxin composed of a mouse monoclonal antibody Fib75 and the ribosome-inactivating protein alpha-sarcin from *Aspergillus giganteus*. Eur J Biochem 1991;196:203–209.

27 Lamy B, Moutaouakil M, Latgé JP, Davies J: Secretion of a potential virulence factor, a fungal ribonucleotoxin, during human aspergillosis infections. Mol Microbiol 1991;5:1811–1815.

28 Paris S, Monod M, Diaquin M, Lamy B, Arruda LK, Punt PJ, Latgé JP: A transformant of *Aspergillus fumigatus* deficient in the antigenic cytotoxin ASPF1. FEMS Microbiol Lett 1993;111: 31–36.

29 Müllbacher A, Eichner RD: Immunosuppression in vitro by a metabolite of a human pathogenic fungus. Proc Natl Acad Sci USA 1984;81:3835–3837.

30 Sutton P, Waring P, Müllbacher A: Exacerbation of invasive aspergillosis by the immunosuppressive fungal metabolite, gliotoxin. Immunol Cell Biol 1996;74:318–322.

31 Jaton-Ogay K, Suter M, Crameri R, Falchetto R, Fatih A, Monod M: Nucleotide sequence of a genomic and a cDNA clone encoding an extracellular alkaline protease of *Aspergillus fumigatus*. FEMS Microbiol Lett 1992;92:163–168.

32 Moutaouakil M, Monod M, Prévost MC, Bouchara JP, Paris S, Latgé JP: Identification of the 33-kDa alkaline protease of *Aspergillus fumigatus* in vitro and in vivo. J Med Microbiol 1993;39: 393–399.

33 Monod M, Paris S, Sanglard D, Jaton-Ogay K, Bille J, Latgé JP: Isolation and characterization of a secreted metalloprotease of *Aspergillus fumigatus*. Infect Immun 1993;61:4099–4104.

34 Jaton-Ogay K, Paris S, Huerre M, Quadroni M, Falchetto R, Togni G, Latgé JP, Monod M: Cloning and disruption of the gene encoding an extracellular metalloprotease of *Aspergillus fumigatus*. Mol Microbiol 1994;14:917–928.

35 Ibrahim-Granet O, D'Enfert C: The *Aspergillus fumigatus mepB* gene encodes an 82 kD intracellular metalloproteinase structurally related to mammalian thimet oligopeptidase. Microbiology 1997;143: 2247–2253.

36 Ramesh MV, Sirakova TD, Kolattukudy PE: Cloning and characterization of the cDNAs and genes (mep20) encoding homologous metalloproteinases from *Aspergillus flavus* and *A. fumigatus*. Gene 1995;165:121–125.

37 Reichard U, Monod M, Rüchel R: Molecular cloning and sequencing of the gene encoding an extracellular aspartic proteinase from *Aspergillus fumigatus*. FEMS Microbiol Lett 1995;130:69–74.

38 Reichard U, Monod M, Odds F, Rüchel R: Virulence of an aspergillopepsin-deficient mutant of *Aspergillus fumigatus* and evidence for another aspartic proteinase linked to the fungal cell wall. J Med Vet Mycol 1997;35:189–195.

39 Beauvais A, Monod M, Wyniger J, Debeaupuis JP, Grouzmann E, Brakch N, Svab J, Hovanessian AG, Latgé JP: Dipeptidyl-peptidase IV secreted by *Aspergillus fumigatus*, a fungus pathogenic to humans. Infect Immun 1997;67:3042–3047.

40 Kobayashi H, Debeaupuis JP, Bouchara JP, Latgé JP: An 88-kilodalton antigen secreted by *Aspergillus fumigatus*. Infect Immun 1993;61:4767–4771.

41 Holdom MD, Hay RJ, Hamilton AJ: Purification, N-terminal amino acid sequence and partial characterization of a Cu,Zn superoxide dismutase from the pathogenic fungus *Aspergillus fumigatus*. Free Radical Res 1995;22:519–531.

42 Hamilton AJ, Holdom MD, Hay RJ: Specific recognition of purified Cu,Zn superoxide dismutase from *Aspergillus fumigatus* by immune human sera. J Clin Microbiol 1995;33:495–496.

43 Hearn VM, Wilson EV, Mackenzie DWR: Analysis of *Aspergillus fumigatus* catalases possessing antigenic activity. J Med Microbiol 1992;36:61–67.

44 Lopez-Medrano R, Ovejero MC, Calera JA, Puente P, Leal F: An immunodominant 90-kilodalton *Aspergillus fumigatus* antigen is the subunit of a catalase. Infect Immun 1995;63:4774–4780.

45 Lopez-Medrano R, Ovejero MC, Calera JA, Puente P, Leal F: Immunoblotting patterns in the serodiagnosis of aspergilloma: Antibody response to the 90 kD *Aspergillus fumigatus* antigen. Eur J Clin Microbiol Infect Dis 1996;15:146–152.

46 Haynes KA, Latgé JP, Rogers TR: Detection of *Aspergillus* antigens associated with invasive infection. J Clin Microbiol 1990;28:2040–2044.

47 Samuelsen H, Karlsson-Borga A, Paulsen BS, Wold JK, Rolfsen W: Purification of a 20-kD allergen from *Aspergillus fumigatus.* Allergy 1991;46:115–124.
48 Haynes KA, Tuinstra P, Hughes TA, Wijnands LM, Rogers TR, Allen AK: Purification and characterization of a 93 kD *Aspergillus fumigatus* antigen with diagnostic potential. J Med Vet Mycol 1996;34:421–426.
49 Taylor ML, Longbottom JL: Partial characterization of rapidly released antigenic/allergenic component (Ag5) of *Aspergillus fumigatus.* J Allergy Clin Immunol 1988;81:548–557.
50 Calera JA, Ovejero MC, Lopez-Medrano L, Segurado M, Puente P, Leal F: Characterization of the *Aspergillus nidulans aspnd1* gene demonstrates that the ASPND1 antigen, which it encodes, and several *Aspergillus fumigatus* immunodominant antigens belong to the same family. Infect Immun 1997;65:1335–1344.
51 Teshima R, Ikebuchi H, Sawada J, Miyachi S, Kitani S, Iwama M, Irie M, Ichinoe M, Terao T: Isolation and characterization of a major allergenic component (gp 55) of *Aspergillus fumigatus.* J Allergy Clin Immunol 1993;92:698–706.
52 Piechura JE, Huang CJ, Cohen SH, Kidd JM, Kurup VP, Calvanico NJ: Antigens of *Aspergillus fumigatus.* II. Electrophoretic and clinical studies. Immunology 1983;49:657–665.
53 Fratamico PM, Buckley HR: Identification and characterization of an immunodominant 58-kilodalton antigen of *Aspergillus fumigatus* recognized by sera of patients with invasive aspergillosis. Infect Immun 1991;59:309–315.
54 KurupVP, Ramasamy M, Greenberger PA, Fink JN: Isolation and characterization of a relevant *Aspergillus fumigatus* antigen with IgG- and IgE-binding activity. Intl Archs Allergy Appl Immunol 1988;86:176–182.
55 Little SA, Warner JO: Improved diagnosis of allergic bronchopulmonary aspergillosis with GP66 (formerly antigen 7) of *Aspergillus fumigatus* for specific IgE detection. J Allergy Clin Immunol 1996;98:55–63.
56 Latgé JP, Kobayashi H, Debeaupuis JP, Diaquin M, Sarfati J, Wieruszeski JM, Parra E, Bouchara JP, Fournet B: Chemical and immunological characterization of the extracellular galactomannan secreted by *Aspergillus fumigatus.* Infect Immun 1994;62:5424–5433.
57 Azuma I, Kimura H, Hirao F, Tsubura E, Yamamura Y, Misaki A: Biochemical and immunological studies on *Aspergillus.* III. Chemical and immunological properties of glycopeptide obtained from *Aspergillus fumigatus.* Jpn J Microbiol 1971;15:237–246.
58 Barreto-Bergter E, Gorin PAJ, Travassos LR: Cell constituents of mycelia and conidia of *Aspergillus fumigatus.* Carbohydr Res 1981;95:205–218.
59 Mischnick P, De Ruiter GA: Application of reductive cleavage in the structural investigation of the antigenic polysaccharides of *Aspergillus fumigatus* and *Penicillium digitatum* with respect to the determination of the ring size of the galactose moieties. Carbohydr Res 1994;23:5–12.
60 Reiss E, Lehmann PF: Galactomannan antigenemia in invasive aspergillosis. Infect Immun 1979; 25:357–365.
61 Barnes RA: *Aspergillus* infection: Does serodiagnostic works? Serodiagn Immunother Infect Dis 1993;5:135–138.
62 Matthews R, Burnie JP, Fox A, Tabaqchali S: Immunoblot analysis of serological responses in invasive aspergillosis. J Clin Pathol 1985;38:1300–1303.
63 Rogers TR, Haynes KA, Barnes RA: Value of antigen detection in predicting invasive pulmonary aspergillosis. Lancet 1990;336:1210–1213.
64 Young RC, Bennett JE: Invasive aspergillosis: Absence of detectable antibody response. Am Rev Respir Dis 1971;104:710–716.
65 Stynen D, Goris A, Sarfati J, Latgé JP: A new sensitive sandwich enzyme-linked immunosorbent assay to detect galactofuran in patients with invasive aspergillosis. J Clin Microbiol 1995; 33:497–500.
66 Stynen D, Sarfati J, Goris A, Prévost MC, Lesourd M, Kamphuis H, Darras V, Latgé JP: Rat monoclonal antibodies against *Aspergillus* galactomannan. Infect Immun 1992;60:2237–2245.
67 Rohrlich P, Sarfati J, Mariani P, Duval P, Carol A, Saint-Martin C, Bingen E, Latgé JP, Vilmer E: Prospective sandwich ELISA galactomannan assay: Early predictive value and clinical use in invasive aspergillosis. Pediatr Infect Dis J 1996;15:321–327.

68 Sulahian A, Tabouret M, Ribaud P, Sarfati J, Gluckman E, Latgé JP, Derouin F: Comparison of an enzyme immunoassay and latex agglutination test for detection of galactomannan in the diagnosis of invasive aspergillosis. Eur J Clin Microbiol Infect Dis 1996;15:139–145.

69 Obayashi T, Yoshida M, Mori T, Goto H, Yasuoka A, Iwasaki H, Teshima H, Kohno S, Horiuchi A, Ito A, Yamaguchi H, Shimada K, Kawai T: Plasma (1-3)-β-*D*-glucan measurement in diagnosis of invasive deep mycosis and fungal febrile episodes. Lancet 1995;345:17–20.

70 Melchers WJG, Verweij PE, van den Hurk P, van Belkum A, De Pauw BE, Hoogkamp-Korstanje JAA, Meis JFGM: General primer-mediated PCR for detection of *Aspergillus* species. J Clin Microbiol 1994;32:1710–1717.

71 Tang CM, Holden DW, Aufauvre-Brown A, Cohen J: The detection of *Aspergillus* spp by the polymerase chain reaction and its evaluation in bronchoalveolar lavage fluid. Am Rev Resp Dis 1993;148:1313–1317.

72 Bart-Delabesse E, Marmorat-Khuong A, Costa JM, Dubreuil-Lemaire ML, Bretagne S: Detection of *Aspergillus* DNA in bronchoalveolar lavage fluid of AIDS patients by polymerase chain reaction. Eur J Clin Microbiol Infect Dis 1996;15:24–25.

73 Verweij PE, Meis JFGM, van den Hurk P, De Pauw BE, Hoogkamp-Korstanje JAA, Melchers WJG: Polymerase chain reaction as a diagnostic tool for invasive aspergillosis: Evaluation in bronchoalveolar lavage fluid from low risk patients. Serodign Immunother Infect Dis 1994;6: 203–208.

74 Verweij PE, Latgé JP, Rijs AJMM, Melchers WJG, De Pauw BE, Hoogkamp-Korstanje JAA, Meis JFGM: Comparison of antigen detection and PCR assay in BAL fluid for diagnosing invasive pulmonary aspergillosis in immunocompromised patients. J Clin Microbiol 1995;33:3150–3153.

75 Fridkin SK, Jarvis WR: Epidemiology of nosocomial fungal infections. Clin Microbiol Rev 1996; 9:499–511.

76 Wald A, Leisenring W, Van Burik JA, Bowden RA: Epidemiology of *Aspergillus* infections in a large cohort of patients undergoing bone marrow transplantation. J Infect Dis 1997;175:1459–1466.

77 Rodriguez E, De Meeus T, Mallié M, Renaud F, Symoens F, Mondon P, Piens MA, Lebeau B, Viviani MA, Grillot R, Nolard N, Chapuis F, Tortorano AM, Bastide JM: Multicentric epidemiological study of *Aspergillus fumigatus* isolates by multilocus enzyme electrophoresis. J Clin Microbiol 1996;34:2559–2568.

78 Symoens F, Viviani MA, Nolard N: Typing by immunoblot of *Aspergillus fumigatus* from nosocomial infections. Mycoses 1993;36:229–237.

79 Girardin H, Latgé JP, Srikantha T, Morrow B, Soll DR: Development of DNA probes to fingerprinting *Aspergillus fumigatus*. J Clin Microbiol 1993;31:1547–1554.

80 Neuvéglise C, Sarfati J, Latgé JP, Paris S: *Afut*1, a retrotransposon-like element from *Aspergillus fumigatus*. Nucl Acids Res 1996;24:1428–1434.

81 Anderson MJ, Gull K, Denning DW: Molecular typing by random amplification of polymorphic DNA and M13 southern hybridization of related paired isolates of *Aspergillus fumigatus*. J Clin Microbiol 1996;34:87–93.

82 Aufauvre-Brown A, Cohen J, Holden DW: Use of randomly amplified polymorphic DNA markers to distinguish isolates of *Aspergillus fumigatus*. J Clin Microbiol 1992;30:2991–2993.

83 van Belkum A, Quint WGV, De Pauw BE, Melchers WJG, Meis JF: Typing of *Aspergillus* species and *Aspergillus fumigatus* isolates by interrepeat polymerase chain reaction. J Clin Microbiol 1993; 31:2502–2505.

84 Loudon KW, Coke AP, Burnie JP: 'Pseudoclusters' and typing by random amplification of polymorphic DNA of *Aspergillus fumigatus*. J Clin Pathol 1995;48:183–184.

85 Field D, Eggert L, MetzgarD, Rose R, Wills C: Use of polymorphic short and clustered coding-region microsatellites to distinguish strains of *Candida albicans*. FEMS Immunol Med Mic 1996; 15:73–79.

86 Bart-Delabesse E, Bretagne S: Polymorphic microsatellite markers of *Aspergillus fumigatus* studied using PCR. 13th Congr of ISHAM, Salsomaggiore Terme, Parma, 1997, p 230.

87 Lin DM, Lehmann PF, Hamory BH, Padhye AA, Durry E, Pinner RW, Lasker BA: Comparison of three typing methods for clinical and environmental isolates of *Aspergillus fumigatus*. J Clin Microbiol 1995;33:1596–1601.

88 Verweij PE, Meis JFGM, Sarfati J, Hoogkamp-Korstanje JAA, Latgé JP, Melchers WJG: Genotypic characterization of sequential *Aspergillus fumigatus* isolates from patients with cystic fibrosis. J Clin Microbiol 1996;34:2595–2597.

89 Leenders A, van Belkum A, Janssen S, De Marie S, Kluytmans J, Wielenga J, Lowenberg B, Verbrugh H: Molecular epidemiology of apparent outbreak of invasive aspergillosis in a hematology ward. J Clin Microbiol 1996;34:345–351.

90 Loudon KW, Burnie JP, Coke AP, Matthews RC: Application of polymerase chain reaction to fingerprinting *Aspergillus fumigatus* by random amplification of polymorphic DNA. J Clin Microbiol 1993;31:1117–1121.

91 Chazalet V, Debeaupuis JP, Sarfati J, Lortholary J, Ribaud P, Shah P, Cornet M, Vu Thien H, Gluckman E, Brücker G, Latgé JP: Molecular typing of environmental and patient isolates of *Aspergillus fumigatus* from various hospital settings. J Clin Microbiol 1998;36:1494–1500.

92 Girardin H, Sarfati J, Traoré F, Dupouy-Camet J, Derouin F, Latgé JP: Molecular epidemiology of nosocomial invasive aspergillosis. J Clin Microbiol 1994;32:684–690.

93 Cenci E, Perito S, Enssle KH, Mosci P, Latgé JP, Romani L, Bistoni F: Th1 and Th2 cytokines in mice with invasive aspergillosis. Infect Immun 1997;65:564–570.

Dr. Jean-Paul Latgé, Laboratoire des Aspergillus, Institut Pasteur,
25, rue Dr. Roux, F–75015 Paris (France),
Fax +33 1 40 61 34 19, E-Mail jplatge@pasteur.fr

Brakhage AA, Jahn B, Schmidt A (eds): *Aspergillus fumigatus*.
Contrib Microbiol. Basel, Karger, 1999, vol 2, pp 88–104

Laboratory Diagnosis of *Aspergillus fumigatus*-Associated Diseases

Reinhard Kappe, Dagmar Rimek

Hygiene-Institut, Ruprecht-Karls-Universität, Heidelberg, Deutschland

During the past 15 years, invasive aspergillosis (IA) has become the main cause of death in patients with acute leukemia and liver transplantation [1, 2]. Therefore, the clinical constellation of fever unresponsive to optimal antibacterial therapy in the presence of granulocytopenia or immunosuppressive therapy, especially with new pulmonary infiltrates, should lead to immediate consideration of performance of diagnostic procedures [3] and empirical use of amphotericin B therapy.

Specimens from the Respiratory Tract

The result of any diagnostic procedure largely depends on the quality of the clinical material available for specific examination in the mycological laboratory. The diagnostic efficiency which may be achieved with various clinical samples from the respiratory tract is presented in table 1. Early IA does not necessarily result in productive cough yielding sputum. As there is hardly any colonization with *Aspergillus* sp. of the upper respiratory tract, even in patients with IA, swabs and washings from the nasopharyngeal or oropharyngeal cavities including saliva are generally inadequate materials, although exceptions have been described [4]. Production of sputum is a sign of pulmonary disease, and the first sputum should be examined for *Aspergillus* by all means. However, noninvasively taken respiratory tract materials (e.g. sputum, induced sputum, tracheal and bronchial secretions and washings) yield disappointingly low overall diagnostic sensitivities [5–7]. The specificity is also limited due to possible colonization of the lower respiratory tract of high-risk patients with *Aspergillus* [8]. Transbronchial biopsy specimens add little to findings of examination of bronchoalveolar lavage and brushing speci-

Table 1. Diagnostic efficiency of specimens from the respiratory tract for invasive pulmonary aspergillosis

Specimen	Efficiency %*
Oral swab	2
Nasopharyngeal swab	5
Mouth washing	5
Saliva	5
Sputum	30
Tracheal secretion	35
Bronchial secretion	40
Bronchial washing	50
Transbronchial biopsy	60
Transthoracic biopsy	70
Open lung biopsy	80

* Estimated for all laboratory procedures.

mens. Even percutaneous transthoracic biopsy and open lung biopsy have limited value in patients with acute leukemia, especially with regard to the outcome of the patient. However, as these procedures provide a means for establishment of a definitive diagnosis by finding tissue invasion, they should be performed as early as possible in the course of any case of pulmonary infection with possible fungal etiology.

Macroscopical Examination of Specimens from the Respiratory Tract

A clinical sample (sputum, tracheal or bronchial secretion or washing) from a part of the lower respiratory tract, which is affected by invasive growth of *Aspergillus*, is likely to be thick, purulent and/or blood-tinged. Sometimes necrotic tissue including fungal elements is expectorated. The samples may contain semisolid, whitish grains comparable to those commonly found in certain cases of subcutaneous mycetoma. If available, these grain-like constituents should be selected for microscopical and cultural examination [9].

Microscopical Examination

Wet Mount

For direct examination, grain-like constituents from respiratory tract specimens can be mounted in a drop of 10% potassium hydroxide on a slide and crushed under a coverglass, as it has been described for grains of mycetomas [10, 11]. In some cases the size and septation of the hyaline hyphae, along with other characteristic morphologic features like dichotomously branched,

1

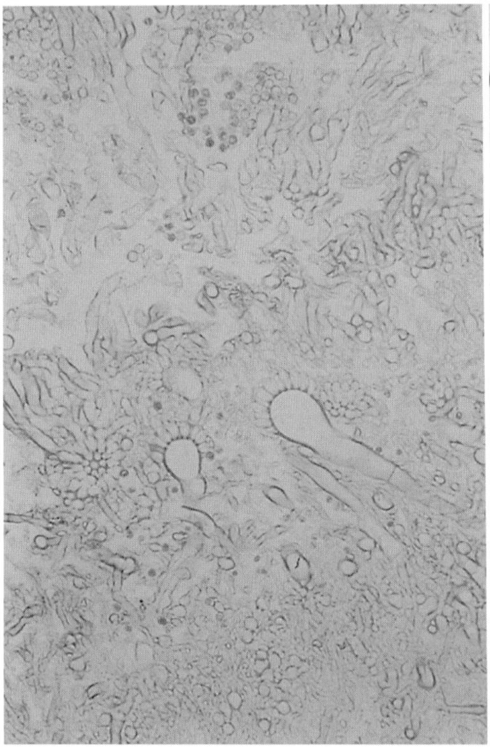

2a

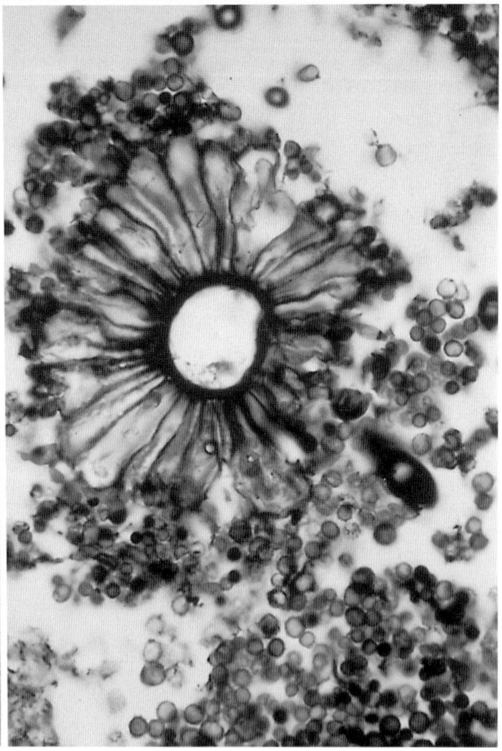

2b

Fig. 1. Wet mount of bronchial lavage fluid from a 45-year-old woman with acute myelogenous leukemia, who died four days later from invasive pulmonary aspergillosis. Septate, hyaline, dichotomously branched, characteristically gnarled hyphae allow an immediate diagnosis. *A. fumigatus* was cultured from this sample. Original magnification × 100.

gnarled hyphae, may allow an immediate diagnosis [9] (fig. 1). The presence of conidial structures indicates that *Aspergillus* is colonizing an area in contact with air. The conidial structures of the two most common species causing cavitary aspergillosis, *Aspergillus fumigatus* and *Aspergillus niger*, are distinct enough for positive identification without necessitating cultural isolation [10] (fig. 2).

Calcofluor Fluorescent Chitin Stain

This is the first choice for staining of clinical material for the microscopic detection of fungi. In addition to the immediate wet mount, the drop of 10% potassium hydroxide contains the fluorescent dye calcofluor white, which specifically stains the chitin in fungal cell walls. This improves the contrast of fungal elements in front of the background [12]. A mercury lamp, a filter with a cutoff of 350 nm, and special lenses are required for the microscope. Identification of fungal elements depends on the examiner's knowledge of fungal morphologic features (fig. 3).

Staining of Histological Sections

Gomori's methenamine silver stain as modified by Grocott [13] is absolutely necessary for the definitive detection of fungi in tissue or the exclusion of fungi, respectively (fig. 4). In immunosuppressed patients, there are dichotomously branched and septate hyphae invading dead or weakened tissue [10, 14]. In these cases, histology without isolation of the fungus will not rule out invasive fusariosis or scedosporiosis (pseudallescheriasis) [15]. This may be achieved, however, by immunoperoxidase staining of tissue sections using specific anti-*Aspergillus* antibodies [16]. Periodic acid Schiff's reagent (PAS) completely fails to stain fungal elements in a significant number of cases. PAS is not a reliable staining method for *Aspergillus*.

Gram Stain

In many cases clinical samples are gram stained only, because a specific search for *Aspergillus* has not been requested by the clinician. Unfortunately, the gram stain does not properly reveal hyphae or conidial structures of Aspergilli.

Fig. 2. a A. fumigatus in a brownish, crumbly piece of tissue, expectorated by a 35-year-old woman, who suffered from aspergillosis at the site of right lower lobectomy because of tuberculosis. Periodic acid Schiff. Original magnification × 100. *b A. niger* in the right upper lobe at autopsy of a 53-year-old woman, who died from invasive aspergillosis after chronic lung disease due to tuberculosis. Gomori's methenamine silver. Original magnification × 250. The presence of conidial structures in both cases indicates that *Aspergillus* was colonizing an area in contact with air. The conidial structures of *A. fumigatus* and *A. niger* are distinct enough for positive identification without necessitating cultural isolation.

3

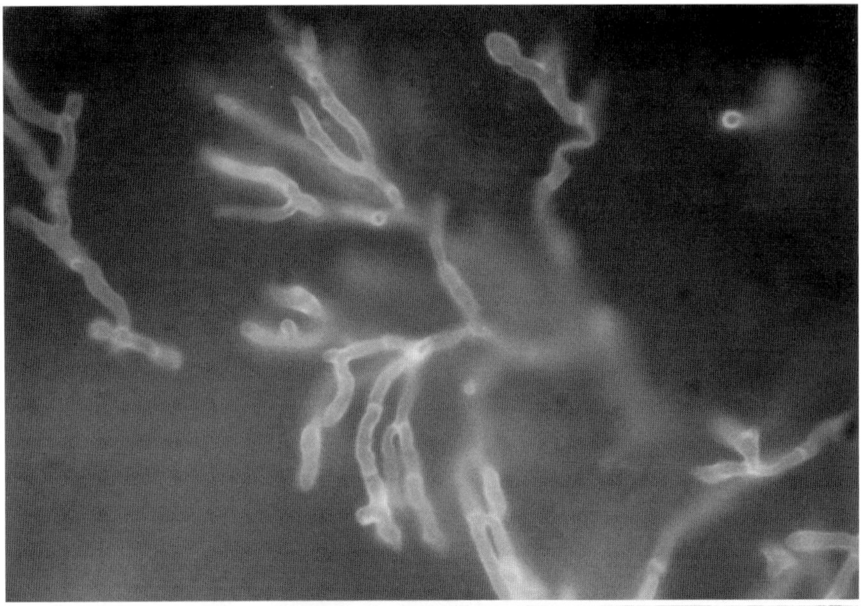

4

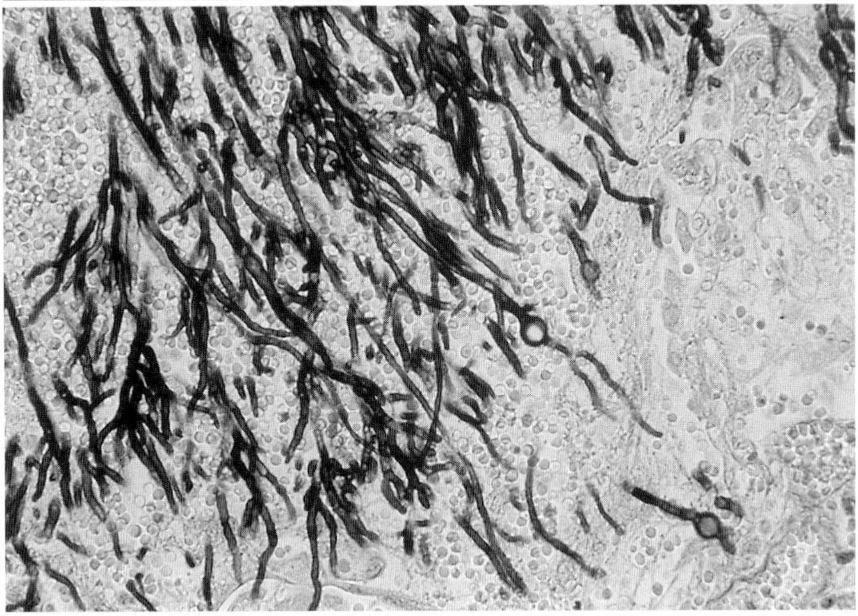

Fig. 3. Calcofluor white stained mount of a 2 × 2 mm selected piece of lung tissue crushed between two slides and mounted in approximately 10 μl of 10% KOH containing 0.1% calcofluor white. The tissue was taken at autopsy from the right lower lobe of the lung of a 45-year-old female with acute myelogenous leukemia who died from invasive pulmonary aspergillosis. Uniform mycelial growth with septae, acute angle branching, and terminal

Table 2. Diagnostic efficiency of laboratory procedures for invasive pulmonary aspergillosis applicable to specimens from the respiratory tract

Procedure	Efficiency %*
Unstained wet mount	15
Gram stain	10
Fluorescence chitin stain	25
Gomori's silver stain	25
Culture solid media	40
Culture broth	35
Antigen detection	25
Antibody detection	not applicable
Nucleic acid amplification	35

* Estimated for all types of respiratory specimen.

Isolation in Culture

Culture is an important diagnostic tool, especially from bronchoalveolar lavage samples [17–19] (table 2). *Aspergillus* species grow well on a variety of conventional agar media including blood or chocolate agar, which are widely used for isolation of bacterial pathogens. The most useful media for isolation from clinical material are modified Sabouraud's glucose (2%) agar and malt extract agar with antibiotics. Most pathogenic and non-pathogenic aspergilli reach optimal growth at a temperature of 30–37 °C (except *Aspergillus versicolor*, which grows best at 22 °C), but sporulation may be more abundant at 30 °C (fig. 5). The conidial structures characteristic of each species (fig. 2) are usually produced within 48–72 h [10, 20, 21].

Liquid or Semiliquid Samples: Respiratory Expectorations, Secretions, and Washings

If there are semisolid, grain-like constituents in the otherwise clear, turbid or viscous liquids, implantation into Sabouraud's agar, possibly in an aerated slant to avoid desiccation, may facilitate the outgrowth of some *Aspergillus*

bulbous swellings are characteristic morphologic features for a number of moulds including *Aspergillus* in tissue. Original magnification × 200.

Fig. 4. Irregular *Aspergillus terreus* hyphae caused a fatal pulmonary bleeding in a 74-year-old patient with acute myeloic leukemia. Bulbous swellings and mycelial constrictions, with globose to oval cells formed at the tips or in an intercalary position, such as that seen here, are typically found in chronic lesions [10, 14]. Gomori's methenamine silver stain. Original magnification × 160.

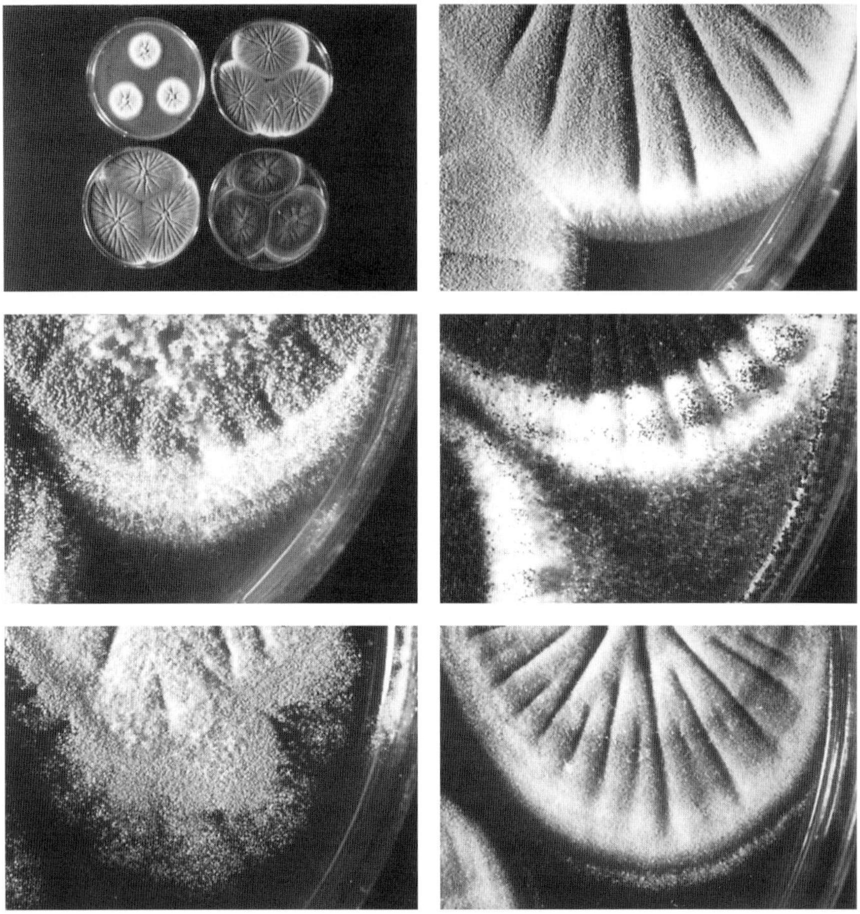

Fig. 5. Cultures of the 5 pathogenic *Aspergillus* species on Sabouraud's glucose (2%) agar. Upper left: *A. fumigatus* 1, 3, 7 and 21 days old. Upper right: *A. fumigatus* 3 days. Middle left: *A. flavus* 7 days. Middle right: *A. niger* 7 days. Lower left: *A. terreus* 7 days. Lower right: *A. nidulans* 7 days.

mycelial fragments with reduced viability due to the host defense or antimycotic chemotherapy. In addition, if there is enough material available for culture, up to 1 ml should be placed into modified Sabouraud's broth containing freshly dissolved penicillin G (50 IU/ml) and streptomycin (50 μg/ml) and incubated at 37 °C for 7 days. Mostly, a pellicle-like fungal thallus is formed on the top of the broth. A low number of irregular hyphae in a necrotic focus, if viable at all, may have physiologically adjusted to a submerged, low oxygen environment. Thus, it may take several days to form a single cotton-like mycelial ball at the bottom of the broth. A subculture of this ball on aerated solid media will reveal typical features.

Biopsy and Autopsy Tissue Samples

The crucial point for successful isolation is to subject the parts of the tissue sample to cultural examination, which contain the fungal elements. Tissue areas, which are infected by *Aspergillus*, are usually altered in colour and consistency, which helps in identification. Mostly they are somewhat whitish and more solid than surrounding unaffected tissue. To create fresh surfaces with possibly viable fungal elements, suspicious nodules should be cut right through the central focus of active fungal growth, to facilitate the outgrowth of mycelia from the tissue into the medium, the surface of the medium or the air. For this purpose, several selected, freshly cut pieces of the native tissue sample are best implanted half into a solid medium like Sabouraud's glucose (2%) agar. To avoid desiccation the agar should be in a slant with a perforated stopper for aeration or a Petri dish, which is incubated in a moist atmosphere or closed with parafilm. The cultures are incubated at 37 °C for 7 days. The combination of partial submersion and partial aeration usually results in growth of a colony within 24 h. Even pleomorphous hyphal fragments with bulbous swellings and constrictions, that are found at the center of chronic lesions, often prove to contain viable elements, and convert to regular morphology. If there is enough tissue available for culture, additional freshly cut pieces should be placed into modified Sabouraud's broth containing freshly prepared penicillin G (50 IU/ml) and streptomycin (50 μg/ml) and incubated at 37 °C for 7 days.

Aspergillus isolates need to be identified to the species level for differentiation of the five pathogenic *Aspergillus* species (fig. 5) from more than ten apathogenic *Aspergillus* species [22, 23]. This is conventionally done by examination of the micromorphology. In addition, species identification by sequence analysis of amplified gene fragments is possible. *A. fumigatus* is by far the most common species causing IPA (fig. 2a). *A. terreus* is emerging as runner up (fig. 4) [24], while previously *A. flavus* has been reported to be second in IA [9]. *Aspergillus niger* is the most common species isolated from swabs of the external auditory canal (fig. 2b). *Aspergillus nidulans* rarely causes IA (fig. 6).

As pathogenic *Aspergillus* species are commonly found in the air, it is necessary to be cautious when examining cultures. A reliable cultural diagnosis can be made only through isolating either several colonies from a single specimen or the same fungus from several specimens. If sufficient material is available, the simultaneous spreading of up to 1 ml (depending on the consistency of the specimen) and 10 μl on separate agar plates allows for differen-

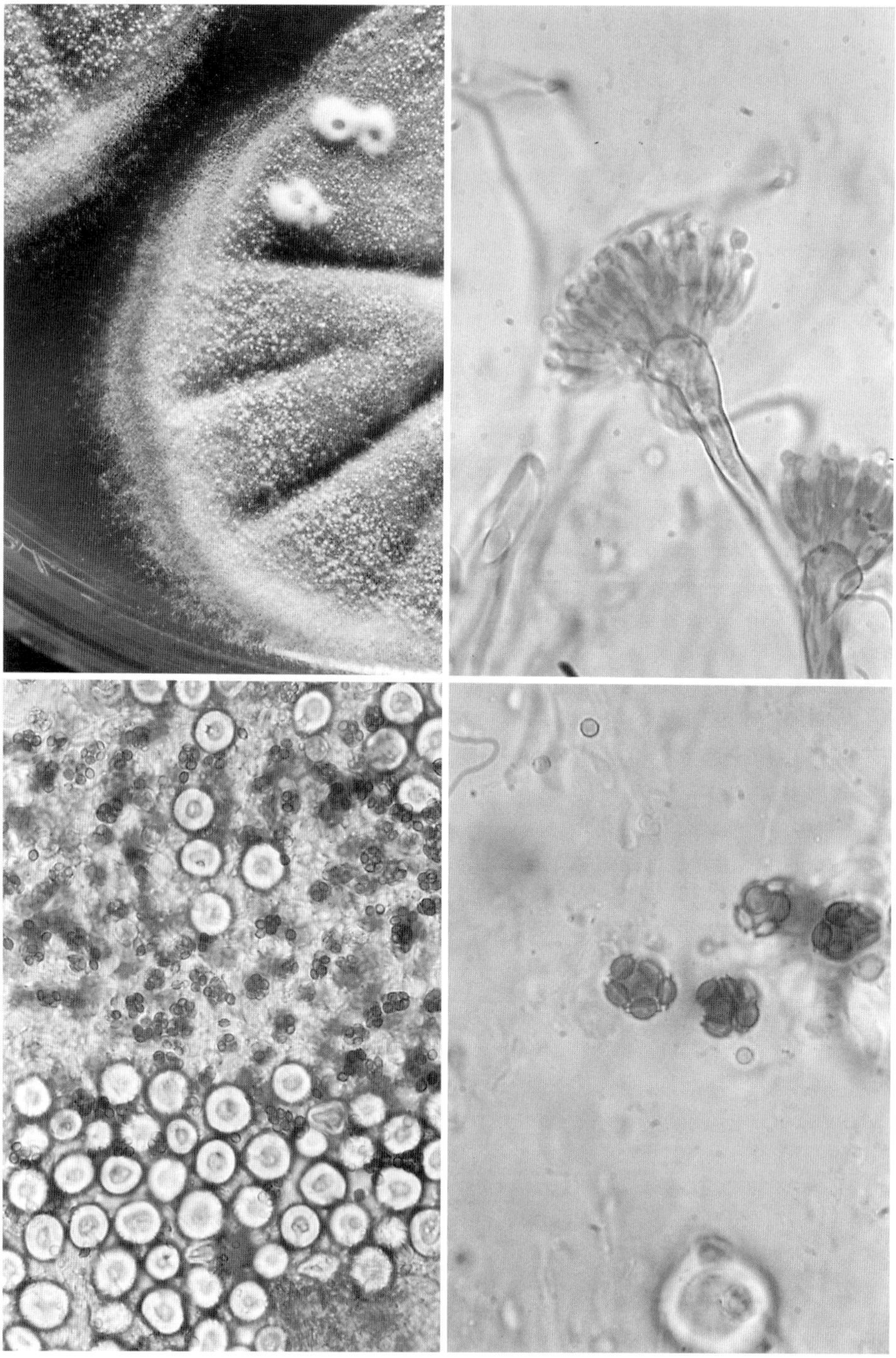

a b c d

Table 3. Cultural procedures recommended for isolation of *Aspergillus* species from a single specimen of the lower respiratory tract

Media[a]	Volume of specimen	Incubation		Reading
		temperature °C	time days	
Agar	1 ml	37	7	daily
Agar	1 ml	28	7	daily
Agar	10 µl	37	7	daily
Agar	10 µl	28	7	daily
Broth	1 ml	37	7	daily
Broth	1 ml	28	7	daily

[a] Modified Sabouraud's media containing 50 IU penicillin G per ml and 50 µg streptomycin per ml.

tiation of an aerial laboratory contamination from a true mould load of the clinical specimen by comparison of the number of colonies on the two plates.

Table 3 summarizes laboratory procedures recommended for optimal recovery and results of cultural isolation of *Aspergillus* sp. from clinical specimens of the deep respiratory tract.

Antigen Detection in Specimens from the Respiratory Tract

A commercial antigen detection kit (ELISA; Platelia® *Aspergillus*; Sanofi Diagnostics Pasteur, Marnes La Coquette, France) is not licensed for clinical samples from the respiratory tract but rather for serum samples. Nevertheless, it has been evaluated for clinical materials like bronchoalveolar lavage and urine in a number of studies [25–28]. The specificity of this antigen detection assay in serum has been found to be 81–100% [29–33]. Cross-reactions of the

Fig. 6. Aspergillus nidulans. a Colony on Sabouraud's glucose agar after 4 weeks at 37 °C. Cleistothecia surrounded by white aerial mycelium are produced on top of the thallus. Diameter of the colony 5 cm. *b* Conidial structures featuring characteristic brown conidiophores and two series of sterigmata originating on the upper half of the vesicle only. Lactophenol cotton blue. Orginal magnification × 320. *c* Cleistothecium dissected with needles and flattened under a coverslip. Globose to subglobose, hyaline, thick-walled, hülle cells are surrounding the ascocarp, which contains numerous asci, each of which contain 8 purple ascospores. Wet mount. Original magnification × 128. *d* Eight double-convex-lens shape, purple ascospores with a furrow flanked by heavy ridges are produced within each ascus. Wet mount. Original magnification × 320.

basic monoclonal antibody to other members of the genus *Aspergillus* are desired (IA by *A. flavus* and *A. terreus* is also detected [34]), and cross-reactions to laboratory contamination [32, 35] or food constituents [36] do not play a significant role. The chemical sensitivity of the system has recently been increased 15-fold (from 15 to 1 ng/ml [37]) by changing the format of the immunoassay from latex agglutination to enzyme immunoassay, while keeping the same rat monoclonal antibody against *Aspergillus* galactomannan antigen as the backbone of the assay [34]. However, for bronchoalveolar lavage, this increase in chemical sensitivity does not result in a significant impact on the clinical sensitivity of antigen detection for the diagnosis of IA, when compared to microscopy, culture, and PCR [27, 28].

Detection of Aspergillus *DNA by PCR*

Several PCR assays have been described, which provide reasonable sensitivities, if examined with purified *Aspergillus* DNA (1 pg to 100 fg) or standardized *Aspergillus* conidiospore suspensions (1–10 spores), and reasonable specificities, if controlled with microorganisms including fungi, bacteria, parasites, and mammalian DNA [38–43]. In contrast, the clinical diagnostic efficiency of PCR assays has been disappointingly low. However, in cases of culture failure due to host response and/or antimycotic chemotherapy, PCR assays have been of limited, yet significant value [27, 44–48].

In summary, as in bacterial infections [49], PCR assays of clinical samples from the lower respiratory tract may complement conventional methods like microscopy and culture for the diagnosis of IA.

Non-Respiratory Tract Samples for Diagnosis of IA

With few exceptions, in overwhelming terminal disseminated IA, blood cultures are generally negative and of no value for the early diagnosis of IA [50]. Urine samples have been recommended for galactomannan antigen detection by latex agglutination, because they appeared to give positive signals earlier and more consistently than serum samples [25, 26, 51]. However, using the enzyme immunoassay, serum has been reported to give better results than urine [37]. Cerebrospinal fluid may yield positive results in antigen detection assays in cases of cerebral dissemination, if there is a connection between the intracerebral *Aspergillus* lesion and the cerebrospinal fluid system.

Cultural Examination of Biopsy Samples

For definitive diagnosis of dissemination, especially in cases of unproven pulmonary IA, biopsy is recommended not only for the lung, but also for suspected *Aspergillus* foci in the skin, the liver, the endocard, the kidneys and other organs. In addition to the histopathological examination, which must include Gomori's methenamine silver stain, the isolation of the fungus in culture should be aimed at. Technical details are described above.

Monitoring of Serum Samples for Circulating *Aspergillus* Antigen, Antibodies, and Nucleic Acid

Galactomannan Antigen Detection

Using the latex agglutination test (LAT, Pastorex® *Aspergillus*; Sanofi Diagnostics Pasteur), circulating antigen was detectable in serum samples only in the very late stages of IA, often shortly before the death of the patients. The LAT has a clinical sensitivity of 36% only [52]. This may be partly due to the rapid clearance of *Aspergillus* galactomannan antigen from the circulation [53]. The commercially available *Aspergillus* galactomannan antigen enzyme immunoassay (EIA, Platelia® *Aspergillus*; Sanofi Diagnostics Pasteur), details of which have been described above, has been improved in sensitivity to result in clinical usefulness [29, 31, 54–56]. The EIA gives positive signals 19–27 days earlier in the course of IA than the latex agglutination test [31, 56]. Serum samples from patients at risk for IA should be examined immediately, and then monitored twice weekly for *Aspergillus* galactomannan antigen by EIA [3].

Antibody Detection

As there appears to exist hardly any background anti-*Aspergillus* antibody level in the normal population, antibody detection, in principal, should be a good means of diagnosing *Aspergillus*-related diseases. Patients with allergic bronchopulmonary aspergillosis and aspergilloma are long known for heavy IgG antibody production, which is easily measureable by using double immunodiffusion, counterimmunoelectrophoresis, hemagglutination or enzyme immunoassays [57].

However, in patients suffering from IA, antibody detection assays ranged disappointingly low in sensitivity (10–39%) [52, 57, 58]. Survivors may produce anti-*Aspergillus* antibodies parallel to the recovery of their white blood cells. Antibody levels fall again with the clearance of the *Aspergillus* from the lung [59]. Therefore, antibody assays are of value for monitoring therapy in patients, who have responded to IA with a rise in antibody [52, 58, 59]. The search for antigens, which are immunodominant in IA, is an ongoing issue [60]. In conclusion, the value of anti-*Aspergillus* antibody detection assays for early diagnosis of IA is very limited.

β-Glucan Detection in Plasma Samples

(1→3)-*β-D*-Glucan is a characteristic major cell-wall component of fungi, and its concentration in normal human plasma is reported to be as low as 10 pg/ml [61]. Recent studies from Japan indicate an elevation of this glucan in plasma of patients suffering from invasive mycoses and fungemias [61–63]. The company ABTI (Worcester, Mass., USA) has developed proprietary reagents for the sensitive detection (100–1,000 pg/ml) of fungal β-glucans in human plasma, urine and other complex biological fluids. However, the clinical efficiency of plasma β-glucan detection for the diagnosis of invasive fungal disease, in particular IA, is unknown until now.

Detection of Circulating Aspergillus *Nucleic Acid*

For cytomegalovirus infection there is clear evidence for both leukocyte-associated DNA and freely circulating DNA in plasma [64]. For fungal PCR assays, sample preparation procedures are crucial. To extract leukocyte-associated microbial nucleic acid, blood needs to be anticoagulated to allow the separation of white blood cells by centrifugation followed by the lysis of the white blood cells. The extraction of fungal DNA includes the use of cell-wall-digesting enzymes like zymolyase and lyticase and of proteinases, all of which must be free from fungal DNA.

For invasive candidosis few authors claim the detection of DNA associated with leukocytes [65–68], whereas the detection of DNA in serum samples has yet to be confirmed [69, 70]. For IA, several authors detected fragments of nucleic acid from *Aspergillus* associated with leukocytes circulating in the bloodstream of patients [68, 71, 72]. One nested PCR assay for the detection of *Aspergillus* DNA in serum of patients with IA has been reported to be superior to antigen detection [73]. A competitive PCR assay for *Aspergillus* DNA in serum was less sensitive than antigen detection [74]. It is assumed, that there is no *Aspergillus* DNA circulating in the bloodstream of healthy individuals or patients suffering from superficial mycoses. Thus, a qualitative PCR assay should be sufficient.

Conclusion

The laboratory diagnosis of IA largely depends on conventional methods like microscopy, including calcofluor white staining, and culture of specimens from the lower respiratory tract. Definitive diagnosis by finding tissue invasion often depends on histological examination of biopsies by Gomori's methenamine silver stain. Antigen detection in serum samples by enzyme immunoassay is recommended for monitoring patients at risk twice weekly. Antibody detection may be useful for monitoring successful therapy and recovery rather than providing an early diagnosis. The detection of *Aspergillus* DNA in bronchoalveolar lavage fluid complements conventional methods. The clinical usefulness of enzyme immunoassays for the detection of circulating β-glucan in plasma and that of PCR assays for the detection of *Aspergillus* DNA in EDTA blood or in serum is under investigation.

References

1 Wald A, Leisenring W, van Burik JA, Bowden RA: Epidemiology of *Aspergillus* infections in a large cohort of patients undergoing bone marrow transplantation. J Infect Dis 1997;175:1459–1466.

2 Kappe R: Fungal pulmonary infections. Curr Opin Infect Dis 1997;10:123–127.

3 Denning DW, Evans EGV, Kibbler CC, Richardson MD, Roberts MM, Rogers TR, Warnock DW, Warren RE: Guidelines for the investigation of invasive fungal infections in haematological malignancy and solid organ transplantation. Eur J Clin Microbiol Infect Dis 1997;16:424–436.

4 Aisner J, Murillo J, Schimpff SC, Steere AC: Invasive aspergillosis in acute leukemia: Correlation with nose culture and antibiotic use. Ann Intern Med 1979;90:4–9.

5 Albelda SM, Talbot GH, Gerson SL, Miller WT, Cassileth PA: Role of fiberoptic bronchoscopy in the diagnosis of invasive pulmonary aspergillosis in patients with acute leukemia. Am J Med 1984;76:1027–1034.

6 Kahn FW, Jones JM, England DM: The role of bronchoalveolar lavage in the diagnosis of invasive pulmonary aspergillosis. Am J Clin Pathol 1986;86:518–523.

7 Levy H, Horak DA, Tegtmeier BR, Yokota SB, Forman SJ: The value of bronchoalveolar lavage and bronchial washings in the diagnosis of invasive pulmonary aspergillosis. Respir Med 1992;86: 243–248.

8 Cahill BC, Hibbs JR, Savik K, Juni BA, Dosland BM, Edin-Stibbe C, Hertz MI: *Aspergillus* airway colonization and invasive disease after lung transplantation. Chest 1997;112:1160–1164.

9 Seibold M: Labordiagnostik der Aspergillose beim abwehrgeschwächten Patienten; in Staib F, Huhn D (eds): Pilzinfektionen bei abwehrgeschwächten Patienten. Berlin, Springer, 1991, pp 17–25.

10 Kwon-Chung KJ, Bennett JE (eds): Medical Mycology. Philadelphia, Lea & Febiger, 1992, pp 201–247.

11 Yuen KY, Woo PCY, Ip MSM, Liang RHS, Chiu EKW, Siau H, Ho PL, Chen FFE, Chan TK: Stage-specific manifestation of mold infections in bone marrow transplant recipients: Risk factors and clinical significance of positive concentrated smears. Clin Infect Dis 1997;25:37–42.

12 Hageage GH Jr, Harrington BJ: Use of calcofluor white in clinical mycology. Lab Med 1984;15: 109–112.

13 Grocott RG: A stain for fungi in tissue sections and smears using Gomori's methenamine-silver nitrate technique. Am J Clin Pathol 1955;25:975–979.

14 Shibuya K, Ando T, Wakayama M, Takaoka M, Uchida K, Naoe S: Pathological spectrum of invasive pulmonary aspergillosis. Jpn J Med Mycol 1997;38:175–181.

15 Guarro J, Gene J: Opportunistic fusarial infections in humans. Eur J Clin Microbiol Infect Dis 1995;14:741–754.

16 Verweij PE, Smedts F, Poot T, Bult P, Hoogkamp-Korstanje JAA, Meis JFGM: Immunoperoxidase staining for identification of *Aspergillus* species in routinely processed tissue sections. J Clin Pathol 1996;49:798–801.

17 Yu VL, Muder RR, Poorsattar A: Significance of isolation of *Aspergillus* from the respiratory tract in diagnosis of invasive pulmonary aspergillosis. Am J Med 1986;81:249–254.

18 McWhinney PHM, Kibbler CC, Hamon MD, Smith OP, Gandhi L, Berger LA, Walesby RK, Hoffbrand AV, Prentice HG: Progress in the diagnosis and management of aspergillosis in bone marrow transplantation: 13 years' experience. Clin Infect Dis 1993;17:397–404.

19 Horvath JA, Dummer S: The use of respiratory-tract cultures in the diagnosis of invasive pulmonary aspergillosis. Am J Med 1996;100:171–178.

20 Salkin IF, McGinnis MR, Cooper CR, Rippon JW, Beneke ES: Current priorities for the clinical mycology laboratory. J Med Vet Mycol 1994;32(suppl 1):309–319.

21 Morris AJ, Byrine TC, Madden JF, Reller LB: Duration of incubation of fungal cultures. J Clin Microbiol 1996;34:1583–1585.

22 Raper KB, Fennell DI (eds): The Genus *Aspergillus*. Baltimore, Williams & Wilkins, 1965.

23 De Hoog GS, Guarro J (eds): Atlas of Clinical Fungi. Baarn & Delft, Centralbureau voor Schimmelcultures, 1995.

24 Kappe R: Trends bei invasiven Mykosen: 81 Aspergillosen in Heidelberg; in Geiss HK (ed): Krankenhaushygiene, Infektiologie, Qualitätssicherung: Paradigmen der modernen Krankenhausmedizin. Pfaffenweiler, Centaurus, 1998, pp 173–189.

25 Rogers TR, Haynes KA, Barnes RA: Value of antigen detection in predicting invasive pulmonary aspergillosis. Lancet 1990;336:1210–1213.

26 Ansorg R, Heintschel von Heinegg E, Rath PM: *Aspergillus* antigenuria compared to antigenemia in bone marrow transplant recipients. Eur J Clin Microbiol Infect Dis 1994;13:582–589.

27 Verweij PE, Latgé, JP, Rijs AJMM, Melchers WJG, de Pauw BE, Hoogkamp-Korstanje JAA, Meis JFGM: Comparison of antigen detection and PCR assay using bronchoalveolar lavage fluid for diagnosing invasive pulmonary aspergillosis in patients receiving treatment for hematological malignancies. J Clin Microbiol 1995;33:3150–3153.

28 Rath PM, Oeffelke R, Müller KD, Ansorg R: Non-value of *Aspergillus* antigen detection in bronchoalveolar lavage fluids of patients undergoing bone marrow transplantation. Mycoses 1996;39: 367–370.

29 Verweij PE, Stynen D, Rijs AJMM, de Pauw BE, Hoogkamp-Korstanje JAA, Meis JFGM: Sandwich enzyme linked immunosorbent assay compared with Pastorex latex agglutination test for diagnosing invasive aspergillosis in immunocompromised patients. J Clin Microbiol 1995;33:1912–1914.

30 Rohrlich P, Sarfati J, Mariani P, Duval M, Carol A, Saint-Martin C, Bingen E, Latgé JP, Vilmer E: Prospective sandwich enzyme-linked immunosorbent assay for serum galactomannan: Early predictive value and clinical use in invasive aspergillosis. Pediatr Infect Dis J 1996;15:232–237.

31 Sulahian A, Tabouret M, Ribaud P, Sarfati J, Gluckman E, Latgé JP, Derouin F: Comparison of an enzyme immunoassay and latex agglutination test for detection of galactomannan in the diagnosis of invasive aspergillosis. Eur J Clin Microbiol Infect Dis 1996;15:139–145.

32 Swanink CMA, Meis JFGM, Rijs AJMM, Donnelly JP, Verweij PE: Specificity of a sandwich enzyme-linked immunosorbent assay for detecting *Aspergillus* galactomannan. J Clin Microbiol 1997;35:257–260.

33 Bretagne S, Marmorat-Khuong A, Kuentz M, Latgé JP, Bart-Delabesse E, Cordonnier C: Serum *Aspergillus* galactomannan antigen testing by sandwich ELISA: Practical use in neutropenic patients. J Infect 1997;35:7–15.

34 Stynen D, Sarfati J, Goris A, Prévost MC, Lesourd M, Kamphuis H, Darras V, Latgé JP: Rat monoclonal antibodies against *Aspergillus* galactomannan. Infect Immun 1992;60:2237–2245.

35 Kappe R, Schulze-Berge A: New cause for false-positive results with the Pastorex *Aspergillus* antigen latex agglutination test. J Clin Microbiol 1993;31:2489–2490.

36 Ansorg R, van den Boom R, Rath PM: Detection of *Aspergillus* galactomannan antigen in foods and antibiotics. Mycoses 1997;40:353–357.

37 Stynen D, Goris A, Sarfati J, Latgé JP: A new sensitive sandwich enzyme-linked immunosorbent assay to detect galactofuran in patients with invasive aspergillosis. J Clin Microbiol 1995;33:497–500.

38 Hopfer RL, Walden P, Setterquist S, Highsmith WE: Detection and differentiation of fungi in clinical specimens using polymerase chain reaction (PCR) amplification and restriction enzyme analysis. J Med Vet Mycol 1993;31:65–75.

39 Reddy LV, Kumar A, Kurup VP: Specific amplification of *Aspergillus fumigatus* DNA by polymerase chain reaction. Mol Cell Probes 1993;7:121–126.

40 Makimura K, Murayama SY, Yamaguchi H: Detection of a wide range of medically important fungi by the polymerase chain reaction. J Med Microbiol 1994;40:358–364.

41 Haynes KA, Westerneng TJ, Fell JW, Moens W: Rapid detection and identification of pathogenic fungi by polymerase chain reaction amplification of large subunit ribosomal DNA. J Med Vet Mycol 1995;33:319–325.

42 Kappe R, Fauser C, Okeke CN, Maiwald M: Universal fungus-specific primer systems and group-specific hybridization oligonucleotides for 18S rDNA. Mycoses 1996;39:25–30.

43 Kappe R, Okeke CN, Fauser C, Maiwald M, Sonntag HG: Molecular probes for the detection of pathogenic fungi in the presence of human tissue. J Med Microbiol 1998;47:811–820.

44 Tang CM, Holden DW, Aufauvre-Brown A, Cohen J: The detection of *Aspergillus* spp. by the polymerase chain reaction and its evaluation in bronchoalveolar lavage fluid. Am Rev Respir Dis 1993;148:1313–1317.

45 Spreadbury C, Holden D, Aufauvre-Brown A, Bainbridge B, Cohen J: Detection of *Aspergillus fumigatus* by polymerase chain reaction. J Clin Microbiol 1993;31:615–621.

46 Melchers WJG, Verweij PE, van den Hurk P, van Belkum A, de Pauw BE, Hoogkamp-Korstanje JAA, Meis JFGM: General primer-mediated PCR for detection of *Aspergillus* species. J Clin Microbiol 1994;32:1710–1717.

47 Makimura K, Murayama SY, Yamaguchi H: Specific detection of *Aspergillus* and *Penicillium* species from respiratory specimens by polymerase chain reaction (PCR). Jpn J Med Sci Biol 1994;47: 141–156.

48 Bretagne S, Costa JM, Marmorat-Khuong A, Poron F, Cordonnier C, Vidaud M, Fleury-Feith J: Detection of *Aspergillus* species DNA in bronchoalveolar lavage samples by competitive PCR. J Clin Microbiol 1995;33:1164–1168.

49 Ieven M, Goossens H: Relevance of nucleic acid amplification techniques for diagnosis of respiratory tract infections in the clinical laboratory. Clin Microbiol Rev 1997;10:242–256.

50 Duthie R, Denning DW: *Aspergillus* fungemia: Report of two cases and review. Clin Infect Dis 1995;20:598–605.

51 Haynes KA, Latgé JP, Rogers TR: Detection of *Aspergillus* antigen associated with invasive infection. J Clin Microbiol 1990;28:2040–2044.

52 Kappe R, Schulze-Berge A, Sonntag HG: Evaluation of eight antibody tests and one antigen test for the diagnosis of invasive aspergillosis. Mycoses 1996;39:13–23.

53 Bennett JE, Friedman MM, Dupont B: Receptor-mediated clearance of *Aspergillus* galactomannan. J Infect Dis 1987;155:1005–1010.

54 Verweij PE, Dompeling EC, Donnelly JP, Schattenberg AVMB, Meis JFGM: Serial monitoring of *Aspergillus* antigen in the early diagnosis of invasive aspergillosis: Preliminary investigations with two examples. Infection 1997;25:86–89.

55 Verweij PE, Erjavec Z, Sluiters W, Goessens W, Rozenberg-Arska M, Debets-Ossenkopp YJ, Guiot HFL, Meis JFGM: Detection of antigen in sera of patients with invasive aspergillosis: Intra- and interlaboratory reproducibility. J Clin Microbiol 1998;36:1612–1616.

56 Herrmann J, Gugel A, Freidank H, Bertz H, Finke J: *Aspergillus*-Antigen-Nachweis: Vergleich eines neuen Sandwich-ELISA mit dem Latex-Agglutinationstest bei Patienten mit histologisch gesicherter invasiver Aspergillose. Mycoses 1998;41(suppl 1):83–85.

57 De Repentigny L: Serodiagnosis of candidiasis, aspergillosis, and cryptococcosis. Clin Infect Dis 1992;14(suppl 1):S11–S22.

58 Manso E, Montillo M, de Sio G, D'Amico S, Discepoli G, Leoni P: Value of antigen and antibody detection in the serological diagnosis of invasive aspergillosis in patients with hematological malignancies. Eur J Clin Microbiol Infect Dis 1994;13:756–760.

59 Tomee JFC, Mannes GPM, van der Bij W, van der Werf TS, de Boer WJ, Koeter GH, Kauffman HF: Serodiagnosis and monitoring of *Aspergillus* infections after lung transplantation. Ann Intern Med 1996;125:197–201.

60 Hearn VM, Pinel C, Blachier S, Ambroise-Thomas P, Grillot R: Specific antibody detection in invasive aspergillosis by analytical isoelectrofocusing and immunoblotting methods. J Clin Microbiol 1995;33:982–986.

61 Obayashi T, Yoshida M, Mori T, Goto H, Yasuoka A, Iwasaki H, Teshima H, Kohno S, Horiuchi A, Ito A, Yamaguchi H, Shimada K, Kawai T: Plasma (1→3)-β-*D*-glucan measurement in diagnosis of invasive deep mycosis and fungal febrile episodes. Lancet 1995;345:17–20.

62 Miyazaki T, Kohno S, Mitsutake K, Maesaki S, Tanaka K, Ishikawa N, Hara K: Plasma (1→3)-β-*D*-glucan and fungal antigenemia in patients with candidemia, aspergillosis, and cryptococcosis. J Clin Microbiol 1995;33:3115–3118.

63 Mori T, Ikemoto H, Matsumura M, Yoshida M, Inada K, Endo S, Ito A, Watanabe S, Yamaguchi H, Mitsuya M, Kodama M, Tani T, Yokota T, Kobayashi T, Kambayashi J, Nakamura T, Masanka T, Teshima H, Yoshinaga T, Kohno S, Hara K, Miyazaki S: Evaluation of plasma (1→3)-β-*D*-glucan measurement by the kinetic turbidimetric *Limulus* test, for the clinical diagnosis of mycotic infections. Eur J Clin Chem Clin Biochem 1997;35:553–560.

64 Boivin G, Handfield J, Toma E, Murray G, Lalonde R, Bergeron MG: Comparative evaluation of the cytomegalovirus DNA load in polymorphonuclear leukocytes and plasma of human immunodeficiency virus-infected subjects. J Infect Dis 1998;177:355–360.

65 Kan VL: Polymerase chain reaction for the diagnosis of candidemia. J Infect Dis 1993;168:779–783.

66 Miyakawa Y, Mabuchi T, Fukazawa Y: New method for detection of *Candida albicans* in human blood by polymerase chain reaction. J Clin Microbiol 1993;31:3344–3347.

67 Holmes AR, Cannon RD, Shepherd MG, Jenkinson HF: Detection of *Candida albicans* and other yeasts in blood by PCR. J Clin Microbiol 1994;32:228–231.

68 Einsele H, Hebart H, Roller G, Löffler J, Rothenhöfer I, Müller CA, Bowden RA, van Burik JA, Engelhard D, Kanz L, Schumacher U: Detection and identification of fungal pathogens in blood by using molecular probes. J Clin Microbiol 1997;35:1353–1360.

69 Chryssanthou E, Andersson B, Petrini B, Löfdahl S, Tollemar J: Detection of *Candida albicans* DNA in serum by polymerase chain reaction. Scand J Infect Dis 1994;26:479–485.

70 Burnie JP, Golbang N, Matthews RC: Semiquantitative polymerase chain reaction enzyme immunoassay for diagnosis of disseminated candidiasis. Eur J Clin Microbiol Infect Dis 1997;16:346–350.

71 Polanco AM, Rodriguez-Tudela JL, Martinez-Suarez JV: Detection of pathogenic fungi in human blood by the polymerase chain reaction. Eur J Clin Microbiol Infect Dis 1995;14:618–621.

72 Van Burik JA, Myerson D, Schreckhise RW, Bowden RA: Panfungal PCR assay for detection of fungal infection in human blood specimens. J Clin Microbiol 1998;36:1169–1175.

73 Yamakami Y, Hashimoto A, Tokimatso I, Nasu M: PCR detection of DNA specific for *Aspergillus* species in serum of patients with invasive aspergillosis. J Clin Microbiol 1996;34:2464–2468.

74 Bretagne S, Costa JM, Bart-Delabesse E, Dhédin N, Rieux C, Cordonnier C: Comparison of serum galactomannan antigen detection and competitive polymerase chain reaction for diagnosing invasive aspergillosis. Clin Infect Dis 1998;26:1407–1412.

Privatdozent Dr. med. Reinhard Kappe, Hygiene-Institut, Ruprecht-Karls-Universität,
Im Neuenheimer Feld 324, D–69120 Heidelberg (Germany)
Tel. +49 6221 567818, Fax +49 6221 565627, E-Mail un691rk@genius.embnet.dkfz-heidelberg.de

Brakhage AA, Jahn B, Schmidt A (eds): *Aspergillus fumigatus*.
Contrib Microbiol. Basel, Karger, 1999, vol 2, pp 105–129

Therapy of *Aspergillus fumigatus*-Related Diseases

Michael Ellis

Division of Infectious Diseases, Department of Medicine, Tawam Hospital, Al-Ain, Abu Dhabi, United Arab Emirates

Aspergillosis is an opportunistic, multifaceted, multisystem disease; its varied clinical manifestations are heavily dependent on the degree of host immunodeficiency and the target organ. High quality clinical efficacy trials of antifungal drugs in the treatment of aspergillosis are sparse. Interpretation of data is plagued by patient heterogeneity, multicentre recruitment, nonuniform entry and response criteria and small patient numbers.

The infective spectrum extends from simple colonisation at one extreme through localized noninvasive diseases such as bronchitis and external otitis to hypersensitivity illness such as allergic bronchopulmonary aspergillosis (ABPA), semiinvasive disease such as some forms of sinusitis, locally invasive forms such as pseudomembranous tracheobronchitis, chronic progressive forms such as chronic necrotising pulmonary aspergillosis to acute fulminant and disseminated forms such as occur in bone marrow transplant recipients at the other extreme. These categories may overlap and in fact often progress from less to more severe forms. This chapter will describe in detail the management of the most challenging and life-threatening entity, namely invasive aspergillosis (IA), particularly in patients with hematologic malignancies or undergoing bone marrow transplantation, and summarizes the salient features of other manifestations.

Acute Invasive Aspergillosis

This entity is virtually confined to patients with notable defects in antifungal defences secondary to reticuloendothelial malignancies, myeloablative and immunosuppressive therapy or who have congenital and/or acquired granulocytic impairment. Several autopsy studies indicate a steep rise in the incidence of invasive fungal infections largely for *Aspergillus* species in patients dying of hematologic and reticuloendothelial cancers, from <10% before 1950 to 40% currently, and with many diagnoses made only at autopsy [1]. Management of IA encompasses several interlinked approaches including local epidemiologics, environmental prophylaxis (e.g. HEPA filters) and building renovation precautions, intra-host primary and/or secondary personal systemic antifungal prophylaxis, early diagnosis and early appropriate treatment. The first two have been addressed by the Centers for Disease Control [1]. Data relating to aspergillosis prophylaxis using oral itraconazole formulations or amphotericin B systemically or aerosolized are limited or conflicting. The earliest but nonspecific sign of IA is antibiotic-resistant neutropenic fever (ARNF). A trial of broad spectrum antibacterials followed by initiation of low dose and dose escalation of amphotericin B, with subsequent dose reduction due to toxicity problems is the usual scenario. The delayed and suboptimal antifungal dosing of amphotericin B over this 7- to 10-day period may have a limited impact on the rapid hyphal proliferation, increase in fungal load and dissemination of *Aspergillus* (fig. 1a, b). A high priority therefore is to make a diagnosis as early as possible. Such attempts have included the use of discriminant score cards based on clinical and radiologic abnormalities. This has reduced the time to initiation of amphotericin B by 4 days compared to conventional clinical judgement [2].

Histologic examination of tissues (fig. 2) will suggest the diagnosis in many cases, but culture is necessary for confirmation. Histology will, in addition, indicate coexisting pathologies, which may require additional treatment. Transbronchial, open lung biopsy or stereotactic biopsy procedures are potentially hazardous in coagulopathic patients. Examination of material obtained by relatively noninvasive techniques from proximal surveillance sites, such as bronchoalveolar lavage (BAL) for pulmonary IA carries a reasonable specificity but low sensitivity. Blood cultures are usually negative in IA.

Attention has therefore focused on serologic testing. Fischer et al. [3] employed serial presymptomatic immunodiffusion-based antibody screening in patients at risk for IA and initiated amphotericin B treatment at *Aspergillus* antibody seroconversion [3]. A twofold improvement in survival to 64% was noted when compared to historical controls. However, this finding must be viewed cautiously in view of the inability of many immunocompromised patients to mount a detectable early antibody response. Identification of immunodominant protein antigens such as the 41-kD antigen of *Aspergillus fumigatus* has permitted the detection of appropriate antibodies in experimental IA [4]. Although some studies indicate as many as 70% patients with IA do produce

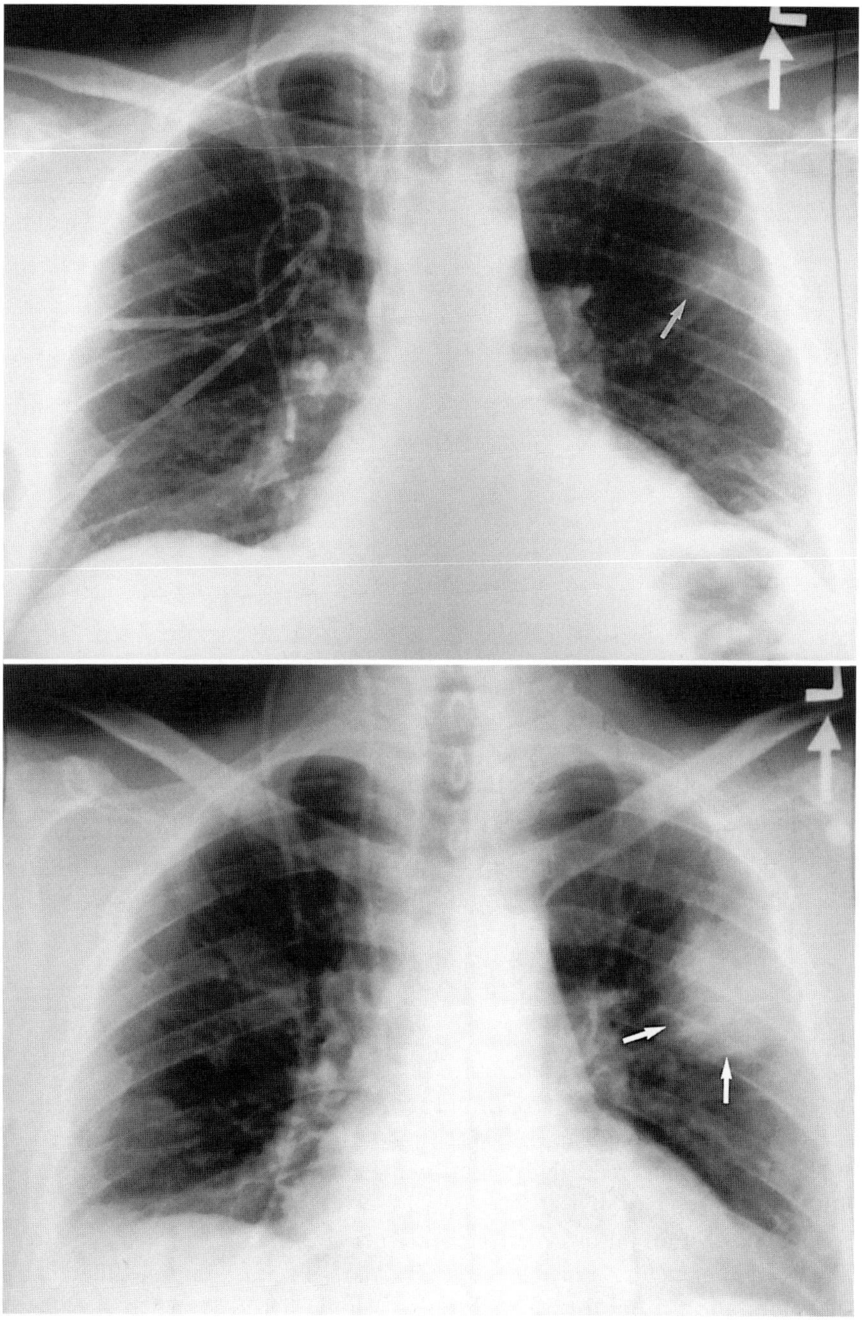

a

b

Fig. 1. a Plain chest radiograph showing indistinct *Aspergillus* lesion in left lung. *b* The same patient 5 days later; note rapid increase in size of lesion.

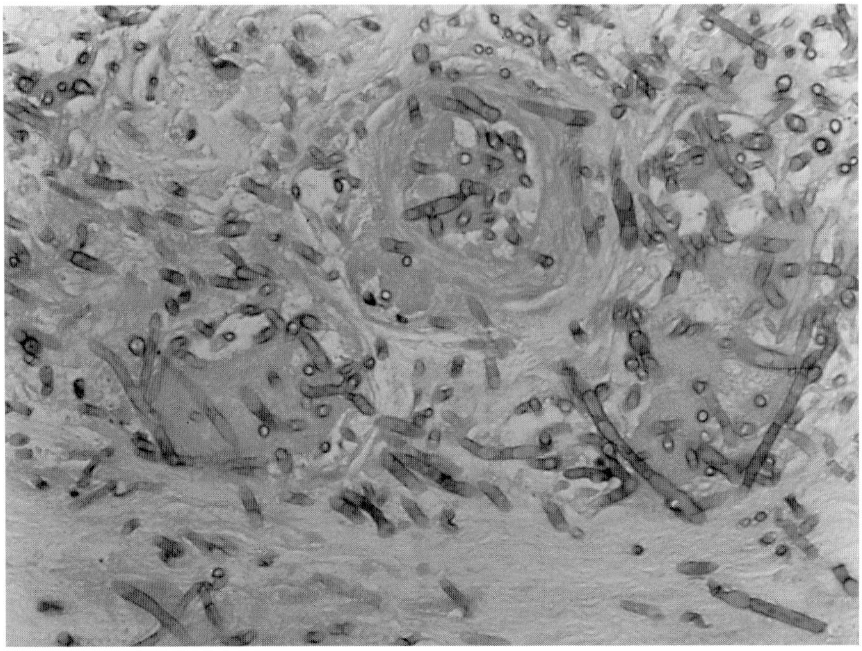

Fig. 2. Septated hyphae with dichotomous branching at 45° angles, showing angioinvasion of *Aspergillus* (tissue biopsy). Grocott-Gomori silver stain. ×400.

detectable antibodies, in general the techniques are insensitive and have not been widely appraised in clinical practice. In contrast, patients with ABPA or aspergillomata do mount antibody responses; serum IgG or *Aspergillus*-specific IgE or IgG and occasionally IgM precipitating antibodies, respectively, are often high and these often correlate with disease activity [5].

Antigen detection methods show greater promise. Sensitivities and specificities of between 75 and 95% (sensitivity) and >90% (specificity) have been reported but are variable depending on the laboratory technique (ELISA/latex), sample type (BAL/serum) and frequency of sampling [6]. Activation of horseshoe crab coagulation factor (factor G) by the (1→3) β-*D*-glucan of fungal cell walls has recently been utilized as the 'G' test with a sensitivity of 90% and a specificity of 100% for invasive mycoses [7].

PCR detection [8] of gene fragment encoding alkaline proteases has encountered interpretation problems of positive results, for example differentiating colonization from disease, and for occasional negative results in patients with presumed aspergillosis. Recent studies using nested PCR to detect a highly conserved region coding for 18s-rRNA in blood samples has proved highly predictive and specific [9]. Other diagnostic tests for oxalate crystals and mannitol production are not in general use currently.

One of the most clinically valuable early diagnostic advances for invasive pulmonary aspergillosis (IPA) has been radiologic [10, 11]. First the acknowledgement that plain chest radiology (CXR) is highly insensitive in early IPA with up to 25% of such patients having a 'normal' CXR. Second, computerized tomography (CT) is often abnormal in the setting of a normal CXR. Third, certain radiographical patterns are highly suggestive, but not pathognomonic, of IPA: Wedge-shaped pleural-based infiltrates, cavitation, air-crescent formation and particularly the precavitary halo sign (fig. 3a, b). The early and serial use of CT scanning in patients with ARNF has dramatically reduced the time to diagnosis and instituting specific treatment for IPA, and this has been associated with improved survival rates of from 30 to 50% in historical controls to >80% [10–12]. It is likely that early CT scanning is only one factor contributing to improved outcome; use of maximally tolerated doses of amphotericin B given earlier, coadministration of 5-flucytosine, adjunctive surgery, and general improvements in patient care have undoubtedly also had a favorable impact.

Amphotericin B

Extensive clinical experience of 35 years dictates that Na-desoxycholate colloid suspension of amphotericin B – conventional amphotericin B (CAB) – has until recently been the drug for therapy of all forms of IA. Many studies compared it with other therapeutic strategies [13]. CAB is fungistatic or fungicidal depending on the tissue concentration achieved and sensitivity of the fungal organism. Mechanisms of action include induction of lethal membrane permeability changes and oxygen radical dependent macrophage stimulation. Toxicity however is formidable and occurs in most patients. This includes systemic reactions (chills, rigors), nephrotoxicity (altered tubular permeability with consecutive negative glomerular filtration feedback), pyuria, albuminuria, hypokalemia, hypomagnesemia and renal tubular acidosis, hematologic compromisation, cardiovascular toxicity, and anaphylaxis. There are also drug interactions with aminoglycosides and cyclosporin which enhance nephrotoxicity, and with granulocyte transfusions or administration of growth factors leading to acute pulmonary distress. Chills and rigors resolve with continued use, or can be modified with slow infusions, premedication with steroids, antihistamines or opiates, and nephrotoxicity possibly diminished by sodium loading or low-dose dopamine. In many patients severe toxicity, particularly renal, continues, leading to temporary dose interruption or even discontinuation of CAB therapy. The ensuing suboptimal treatment negatively

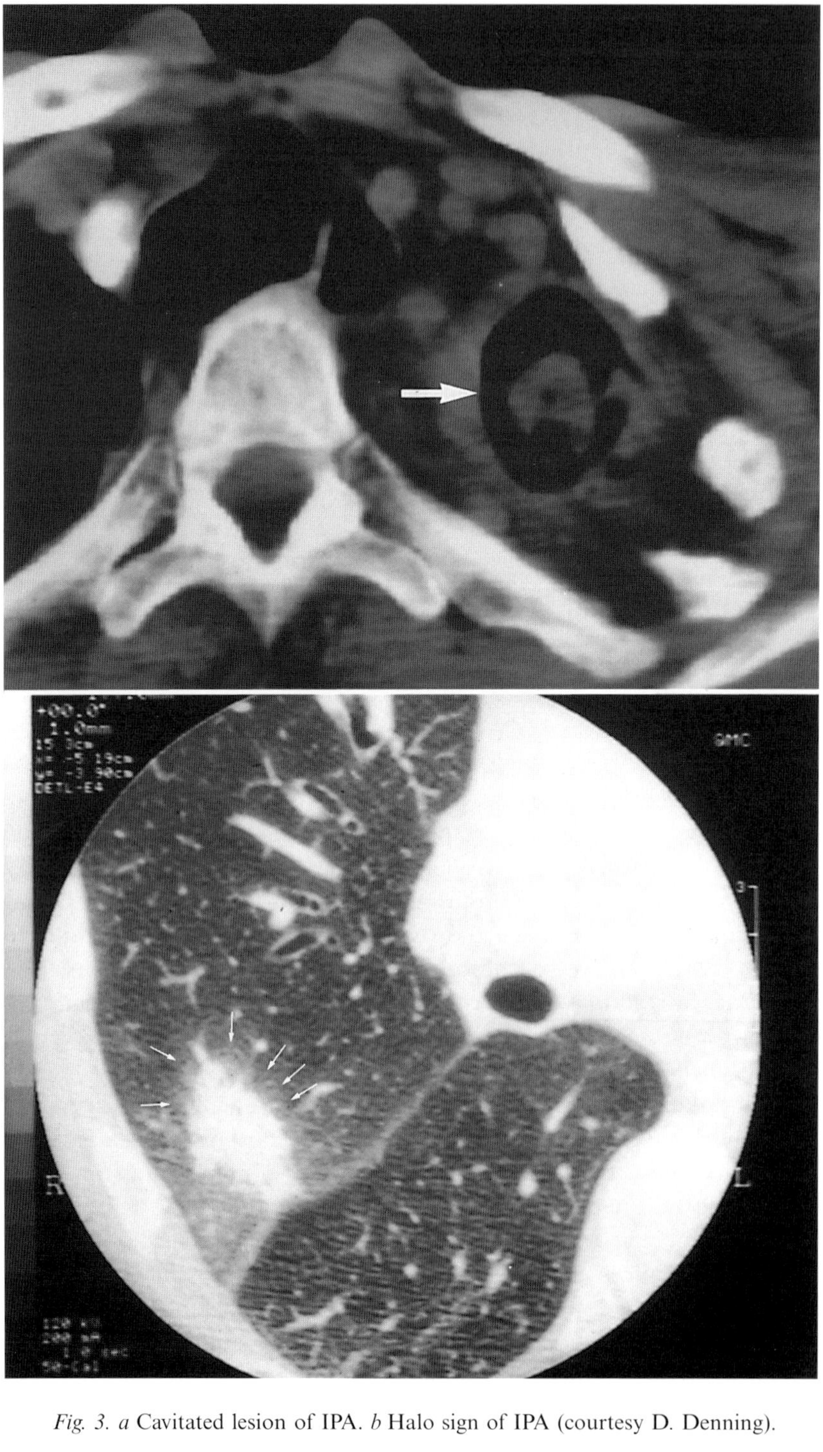

a

b

Fig. 3. *a* Cavitated lesion of IPA. *b* Halo sign of IPA (courtesy D. Denning).

impacts on therapeutic outcome. Crude overall response rates in immunocompromised patients with IPA are around 30%, with an ultimate survivorship even lower, and lowest in patients compromised by HIV infection or with bone marrow transplantation (mortality rates >90%) [14, 15]. Evidence is accruing that patients with renal, heart or lung transplants and IA respond better to therapy (80% response) whilst liver transplant recipients do poorly (20% response) [14]. IA in HIV-infected patients occurs in advanced AIDS patients who have CD4 counts of $<100/mm^3$ often associated with intercurrent disease such as *Pneumocystis carinii* infection, nontuberculous mycobacteriosis, or toxoplasmosis [16]. Response to antifungal treatment is extremely poor in these patients whilst survivors often relapse or succumb to other opportunistic infections.

Animal models indicate that high-dose CAB administered early optimizes survival such as in murine systemic infections caused by *A. fumigatus*. Also, for example in the guinea pig aspergillosis model, survival increased from 17 to 42% when CAB dosage was increased from 0.63 to 2.5 mg/kg [17, 18]. Human clinical observations also suggest this. Burch et al. [19] treated 14 of 15 patients (11 of whom had developed IPA whilst receiving 0.5 mg/kg CAB) with 1–1.5 mg/kg CAB (of whom 10 also received 5FC). Thirteen of the 14 patients survived whilst 1, who continued treatment with low-dose CAB (0.5 mg/kg), died [19]. All 5 patients reported by Ruchlemer et al. [20] and treated with high doses of CAB responded well to therapy [20]. However, data from large prospective randomized studies are desperately needed to scientifically conclude that higher doses of CAB are more beneficial. Despite this limitation, it does seem logical to escalate to the maximum tolerated dose of CAB within 2–3 days of commencing therapy, and to continue until bone marrow recovery occurs and there is clinical and radiologic resolution of IA – in effect this approximates a cumulative dose of 1.5–2 g for an adult.

Nephrotoxicity is the key limiting factor in the use of CAB. A previous observation that CAB toxicity was diminished in the presence of serum low-density lipoproteins led to the development of lipid formulations of amphotericin B, hence providing the potential for administering higher doses of amphotericin B with reduced toxicity. At least 6 different lipid preparations of amphotericin B have undergone clinical trials of which liposomal amphotericin B (AmBisome™), amphotericin B lipid complex (ABLC; Abelcet) and amphotericin B colloidal dispersion (ABCD) have been extensively investigated and are currently marketed. A 'homemade' nonstandardized preparation of CAB with intralipid has been used by some clinicians but is nephrotoxic and of unproved value. Composition, shape, diameter influencing stability, bioavailability and toxicity all vary between the different formulations. These details are presented in depth elsewhere [21]. Substantially reduced nephrotoxicity compared to CAB is well demonstrated for all 3 preparations (AmBisome, ABLC, ABCD), but chills and/or severe systemic reactions appear to remain common with ABCD [22]. Vast experience has accrued with the most well-established preparation – liposomal amphotericin B (AmBisome). This is highly stable and is preferentially taken up by the reticuloendothelial system with resultant low serum levels of free amphotericin B, which may explain the substantial reduction in toxicological side effects. Direct targeting of the fungal cell wall occurs [23]. The subsequent release of amphotericin B into fungal cytoplasm causes cell death in a number of ways, e.g. by membrane permeability changes (e.g. via lipid peroxidation), proton ATP-ase inhibition, and the formation of transmembrane channels [21].

Evidence is beginning to accrue suggesting that the lipid amphotericin B products may have improved efficacy over CAB. Rat and rabbit model data indicate twofold improvements in survival and severalfold improvement over CAB in tissue sterilization and substantially reduced extrapulmonary dissemination of *Aspergillus* species [18, 24, 25]. A large volume of data from single case reports and several open-label retrospective studies in which the lipid preparations were given to patients developing toxicity and/or experiencing progressive IA disease whilst receiving CAB ('salvage' treatment) indicate subsequent responses of up to 60% in patients given liposomal amphotericin B (47% for ABLC and 33% for ABCD, respectively) [26–30]. A meta-analysis of 5 studies in which ABCD given as salvage therapy was compared with historical controls given CAB showed significantly better clinical responses (49 vs. 23%) and survivorship for ABCD [31]. The mean cumulative dosage for ABCD was 6-fold higher than that for CAB and this may have been the reason for the better outcome. There are limitations in the interpretation of such data derived from retrospective studies undertaken under very different conditions from different centers, with different drug dosages, different adjunctive therapies, different neutropenic profiles, and different incidences of IA at autopsy. Careful manipulation of data to correct for these confounders has been done in only one study [31] but bias cannot be eliminated completely. Among 71 studies of *Aspergillus* and/or other mould infections, 33 different sets of criteria were used for definitions of IA, highlighting that even definitions are highly discrepant [32]. Nevertheless, the data taken as a whole do provide a strong hint that amphotericin B lipid formulations might be more efficacious than CAB in treating IA and gives solid support for prospective comparative randomized trials.

Indeed, some comparative studies have already been concluded. EORTC No. 19923 was a pivotal study designed to address the issue of whether 4 mg was more efficacious than 1 mg/kg/day of AmBisome in treating leukemic and bone marrow transplant patients with IA [33]. Of 120 randomized patients 87 with presumptive/definitive IA (ratio 3:1) were eligible for analysis. The clinical and radiologic responses were similar at 50–60% for each arm at the end of the treatment period. Survival for all eligible patients by intention to treat analysis and for subgroups (e.g. leukemia patients alone) was similar for each dose, namely 50% at 2 months and 35–40% at 6 months. Directly attributable IA mortality was 30–35% at 6 months. These findings are unexpected given the results from animal models which provide evidence of dose-response [25, 26]. However two other human studies also failed to demonstrate significantly increased efficacy of a higher dose of AmBisome in treating ARNF [34, 35]. The possible explanations include: (1) the slight differences between the two arms such as neutropenia, intensity of immunosuppression, ratio of presumptive/definitive IA cases although insignificant as single phenomena might cumulatively have obscured any real advantage of the 4-mg dose; (2) there may have been hidden confounders not identified by the statistical techniques; (3) a higher AmBisome dose might be immunosuppressive, or (4) an even higher dose than 4 mg is needed to show any difference. In the small subgroup of 20 patients with definitive IA at randomization, clinical response, IA attributable mortality and survival were all 1.5- to 2-fold higher on 4 mg – a nonsignificant observation however.

Results of some prospective studies comparing AmBisome and CAB in leukemic and BMT patients are summarized in table 1. In patients with IA and in patients with other invasive fungal infections as well as patients with cryptic fungal infection (ARNF) there does appear to be a better response with AmBisome [34–37]. In such a study of almost 800 patients performed by Walsh et al. [36] there was a significant reduction in the emergence of all invasive fungal infections and a trend in reducing IA in patients receiving 3 mg AmBisome compared to CAB. The findings of these studies are extremely important since if the premise that ARNF is the earliest, but nonspecific, sign of invasive fungal disease is accepted, then liposomal amphotericin B does appear to be more effective when administered at a very early stage, in preventing clinically apparent IA. AmBisome was also shown to be superior to CAB in the treatment of neutropenia-associated invasive fungal infections [37].

Suggestions for the Use of Amphotericin B in Invasive Aspergillosis

Consensus opinion is that if CAB is used it should be administered as early as possible in the course of the disease and therapy certainly should not be delayed whilst awaiting histology/culture identification [15]. The dosage

Table 1. AmBisome and CAB relative efficacies

Study	Disease	% responding				% surviving			
		CAB	AMB1	AMB3	AMB4	CAB	AMB1	AMB3	AMB4
Ellis et al. [33] EORTC No. 19923	IA	–	63	–	48	–	58	–	51
Prentice et al. [34]	ARNF	49	58	64		–	–	–	–
		(CAB vs AMB1 *)							
Hann et al. [35]	IA	22	50	67	–	–	–	–	–
Walsh et al. [36]	ARNF	49	–	50	–	–	–	–	–
	IFI	9.3	–	4.7	–	–	–	–	–
		(CAB vs AMB3 *)							
Leenders et al. [37]	IA and IFI	18	–	–	44	38	–	–	22
		(CAB vs AMB4 *)							

IFI = Invasive fungal infections; – = not done. AMB1 = 1 mg/kg/day of AmBisome; AMB3 = 3 mg/kg/day of AmBisome; AMB4 = 4 mg/kg/day of AmBisome.
* $p < 0.05$.

should be 1.0–1.5 mg/kg/day. Switch to a lipid formulation is indicated for intolerable systemic toxicity or severe hematologic and/or renal toxicity. The above are generally accepted guidelines, to date. There is, however, considerable debate surrounding the use of lipid formulations – which compound, timing of administration in the course of the disease (e.g. after amphotericin B toxicity/failure or as primary treatment), optimal dosage and duration of therapy, level dosage or dose escalation and to what ceiling, or initially high dosage followed by decreasing dosage titrated to response. These can only be addressed by well-controlled studies. The options currently therefore are: (a) lipid amphotericin B from the start of treatment; (b) dosage 1 mg/kg/day throughout treatment; (c) initial dose of 1 mg/kg/day then increase to the maximum tolerated (5–10 mg/kg/day or beyond) for failure to respond, and (d) initial dosage of 5–10 mg/kg/day and reduce in case of response.

Human clinical studies have not yet been performed to address these issues. Preliminary evidence suggests an advantage for lipid amphotericin B from the start of treatment. Animal experiments and theoretical considerations as well as isolated case reports suggest that either (b) or (c) (see above) are clinically acceptable alternatives, whilst option (d) is novel. In addition to the administration of CAB or lipid amphotericin B, augmentation of host response should neutropenia persist (see below) in the face of 'optimal' antifungal antimicrobials, with appropriate adjunctive surgery (see below) and empirical addition of other antifungals (see above) should be considered.

Itraconazole

Itraconazole (ITRA) is an alternative treatment for IA [38, 39]. Once again studies have been largely open, salvage design, and of variable and limited quality. Crude efficacy rates are on a par with those for CAB, and again are organ and immunocompromized profile-dependent. An interesting small series of patients with cerebral abscesses giving a cure rate of 25% was reported recently – impressive when mortality would be expected to be >95% in these patients and also lends credence to the better effect of using higher dosages [40]. There are notable problems with ITRA – substantial inter-/intrapatient serum level variability, subtherapeutic MICs in 1/3 of the patients, drug interactions, and drug toxicity especially in the case of preexisting hepatopathy. These relegate the role for ITRA to less-ill patients tolerant of oral medications, in postengraftment, post-CAB-responsive individuals, secondary prophylaxis, in selected patients with severe CAB toxicity, and as adjunctive or alternative therapy in those with progressive disease on amphotericin B treatment. Antagonistic effects between ITRA and antimycotics such as amphotericin B are still not sufficiently evaluated.

Combination Antifungals

Goals of such treatment include augmentation of clinical efficacy, reduced toxicity and/or broadening of the antifungal activity spectrum. Some of these have been achieved in animal models – for 5-fluorocytosine (5FC) + ITRA, CAB + rifampicin (RIFA), CAB + 5FC, but not for CAB + ITRA [17]. In the human clinical arena, there has been little planned evaluation of such an approach. Overall, marginally better responses only for the CAB + 5FC combination compared to CAB alone are seen from open studies [15], but a small randomized study showed no such advantage [41]. Provided serum levels are <100 μg/ml, hematopoiesis inhibition and delayed hematopoietic recovery known to occur in vitro and in vivo, are probably not clinically significant. The theoretical concern that concurrent use of ITRA and CAB is unsound since prevention of CAB binding may occur has backing from in vitro studies where a 3× increase in strains demonstrating antagonism has been shown [42]. In clinical practice ITRA is often used sequentially or for secondary prophylaxis with good efficacy and no obvious antagonism. The CAB + RIFA combination may pose P450 induction problems; and there is virtually no information on ITRA + 5FC or RIFA + ITRA. So in general there is very little evidence to support combination therapies apart from 5FC + CAB in patients with brain disease, fungal endocarditis, or in patients deteriorating on CAB. Even so the majority of patients with IA do receive combination therapy.

Failure of Treatment

Many of the clinical failures with IA are directly linked to the massive host defence failure. It is unclear what role acquired or intrinsic antifungal drug resistance plays, because of lack of reproducible susceptibility testing and poor in vitro/in vivo correlation. Standardized in vitro methods of resistance testing, correlated to clinical outcome are now achievable for infections with some *Candida* species. For *Aspergillus* it is certainly possible to generate reproducible MICs which may correlate with clinical outcome in animal models, but application to the human situation is an urgent need of further evaluation. However, some cases of clinical failure have been correlated with high MICs to *A. fumigatus* [43].

Surgery

Clear indications for adjunctive surgery are listed below.

Endocarditis

There are few survivors reported in the literature as a result of receiving medical therapy alone. Mortality falls to 25–50% and embolic complications are minimized with a combined approach of CAB with/without 5FC and valve replacement [44].

Endopthalmitis, Bone and Joint Disease

These diseases often fail on pharmacological therapy alone and need surgical intervention [15].

Early Bleeding from Cavitary Mycotic Lung Sequestrum (MLS)

Emergency surgery significantly reduces the 25% mortality in patients who are at risk from bleeding, usually at the time of neutrophil recovery and migration [45].

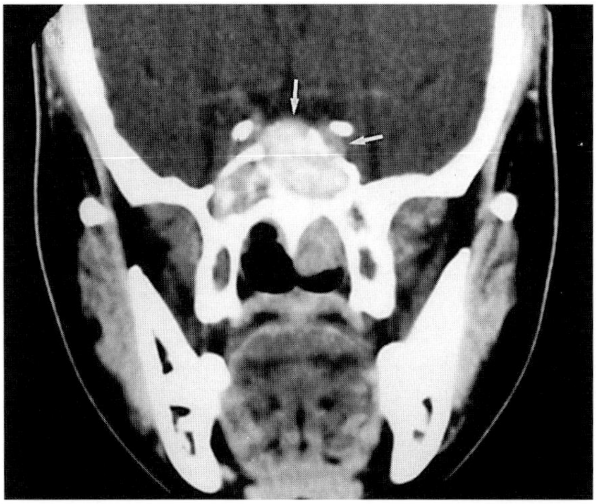

Fig. 4. CT skull: note sinus *Aspergillus* lesion invading brain tissue.

Prevention of Bleeding in Well-Circumscribed IA Masses

Preemptive removal may reduce the risk of hemoptysis – 30% of such patients are thought to be at risk. A few papers support this contention [46, 62]. Although operative morbidity is highly variable it is not usually major, despite the ongoing granulocytopenia. Attributable mortality is unrelated to the procedure with most deaths due to ongoing multiorgan failure and/or sepsis. However, the complication rate in such surgery does appear to be highest for *Aspergillus* lung disease rather than other invasive fungal infections.

Refractory Cases

Benefit is less clearly established. Isolated case reports suggest that judicious surgery in slowly progressive rather than fulminant IA may halt disease progression [47].

Prevention of Future Relapse/Recurrence

This is the most difficult area to disentangle. Removal of surgically accessible foci in patients surviving IPA could prevent the 50% reactivation rate during further immunosuppression. Some papers support this approach in patients who have been followed up for as long as 18 months [46]. However, emergence of *Aspergillus* in extrapulmonary organs may occur. Prolonged follow-up may permit relapse, even in patients who have had adequate initial resection [48]. It is difficult to be sure just how useful surgery is since in practice many patients do also receive antifungal drugs. Large lesions of *Aspergillus* in a protected vascular

Table 2. Sinusitis

Presentation	Treatment
Colonization	none
Local AS	surgery*
Semiinvasive AS	surgery/surgery + antifungals**
Invasive AS	antifungals/antifungals + surgery

* Add antifungals for ethmoid disease or *A. flavus* infection.

** Add antifungals if any immunocompromisation (even minor, e.g. diabetes mellitus).

AS = *Aspergillus* sinusitis.

environment may not benefit from antifungal drugs, but there are studies that do show that antifungal drugs alone can be effective [49].

Sinusitis

It is now recognized that a continuum of disease with overlap and progression exists through the broad groupings of *Aspergillus* sinusitis (AS) (table 2) [50]. Management hinges on assessment of immune status, radiologic delineation of bone and intracranial spread (fig. 4), and histologic invasiveness. Local AS and semiinvasive sinusitis (i.e. bone destruction without demonstrable submucosal hyphae) can be successfully managed with surgery alone. Some clinicians add antifungals for semiinvasiveness though one report indicates that this is unnecessary, except in patients with any degree of immunocompromization (e.g. diabetes mellitus) [51]. A subgroup of patients with ethmoid sinusitis can sometimes develop dissemination and intracranial spread [52]. Another group of patients, apparently otherwise healthy, can react initially with local AS progressing to chronic slowly progressive erosive and destructive granulomatous AS often due to *A. flavus*, with a poor long-term outcome [53]. As many as 80% recur often with intracranial spread and high mortality. These patients fail to respond to CAB. Despite extensive surgical resection using skull-base tumor techniques, the cure rate can be no greater than 10% [52]. The use of ITRA postoperatively has been associated with a higher remission rate (63% at 17 months follow-up) with only 1/22 patients progressing [52, 54]. Long-term ITRA has often been used for 'consolidation' antimycotic treatment in such patients.

In severely immunocompromised patients with acute or fulminant AS, antifungals provide the management template upon which definitive surgery may be performed – but in who and when is debatable [55]. Arguing that surgery is hazardous mainly from bleeding which may be fatal and the associated high mortality rates, some reserve its place for progressive disease on antifungals [15, 55, 56]. The most recent literature appears to lend support for early surgery [57, 58] with careful assessment via functional endoscopy and extensive initial debridement, reserving radical removal for progression, with systemic amphotericin B therapy. Most in this series survived in spite of persistent neutropenia at surgery. This early approach was supported by Kavanagh et al. [57]; the only deaths occurred in patients which were persistently neutropenic postoperatively and were directly related to relentless hematologic disease. HIV-associated sinusitis occurs in only 3% of the patients against a background of coinfection with *Cytomegalovirus, Mycobacterium avium-intracellulare,* and *Pneumocystis carinii,* and profound CD4 leukocytopenia. It is at least as destructive as in patients with hematological malignancies. Response to antifungals and/or surgery is depressing – only 12% of the patients respond [59].

Rare Invasive Aspergillosis Manifestations

Pseudomembranous Necrotizing Bronchial Aspergillosis
This entity affects large airways, with peribronchial invasion. Signs are subtle. Unilateral wheeze may be present. Bronchoscopy is diagnostic. If unrecognized dissemination occurs, the outcome may be fatal [60]. Treatment is bronchoscopic resection and therapy with CAB or ITRA.

Invasive Oral Aspergillosis
Initially, there is marginal gingivitis which under the intense immunosuppressive drive progresses to acute necrotizing ulcerogingivitis. Treatment is with debridement and CAB. Success is determined by neutrophil recovery [61].

Cutaneous Invasive Aspergillosis
This occurs after local implantation e.g. burns, intravenous catheters, but the hematogeneous route is more common in the immunocompromised patient. Rapid necrosis ensues. Surgery and CAB is indicated. Hyperbaric oxygen treatment is experimental [62].

Central Nervous System Invasive Aspergillosis
In the immunocompromised this manifestation of IA carries the highest mortality [15, 40] but in immunocompetent patients survival is normal. Usually, a combined approach with surgery, drainage/removal, and systemic anti-

fungals, namely CAB or high dose ITRA with 5FC is indicated. Local intracavitary instillation may be successful when systemic CAB has failed [64]. Therapeutic dogged persistence is necessary.

Adjunctive Treatments

Granulocytes and Growth Factors

Recovery from IA is heavily dependent on neutrophil recovery – what is the current status with granulocyte transfusions and the use of recombinant growth factors?

White Cell Transfusions

Metaanalysis supports their short-term benefit in bacterial sepsis in patients with cancer but the efficacy in patients with invasive fungal infections is poor [65]. This does not mean that white cell transfusions could not be beneficial since details of transfusion techniques, HLA matching, etc. in previous reports are limited. Current donor harvest technology has improved considerably to substantially increase cell yield, and quality, and reduce apoptosis, and risk of transfusion associated infection. The contribution of white cell transfusions in managing patients with IA should be reassessed.

Human Growth Factors

The effect of these and other cytokines on neutropenia duration and prolonged neutrophil function should theoretically be helpful in IA. In vitro data indicate that maximal antifungal effects are achieved with a combination of an antifungal (e.g. voriconazole), adequate neutrophils, and granulocyte colony-stimulating factor [66]. However, no consistent clinically meaningful effect such as reduction in hospital stay, overall survival, and infection deaths has been demonstrated when they have been used in bacterial sepsis. Granulocyte-monocyte colony-stimulating factor (GMCSF) shows a trend towards improved response in patients with invasive fungal infections but its pulmonary toxicity is problematic [67]. Studies comparing outcome in patients on optimal antifungals versus those who also receive growth factors are in progress. At this stage it would nevertheless seem justified to use colony-stimulating factors in persistently neutropenic patients receiving appropriate treatment for IA. The paper by Cenciu et al. [68] indicating that 70% of mice with IA could be cured with soluble IL-4 receptor treatment, suggests that exploration of the cytokine web could produce some interesting results.

Hyperbaric Oxygen

White-cell-mediated fungal killing is enhanced in a hyperbaric environment. Optimal tissue oxygenation, reduction in edema and acidosis are additional benefits which might also be expected to improve tissue survival and accelerate healing. Hyperbaric oxygen has been successfully used in therapy of zygomycosis [69] but there is little experience in the field of IA. Adjunctive use in some patients, e.g. who have progressive extensive tissue destruction due to IA should be considered, and might prove useful (fig. 5a, b, c).

Intensive Care

The literature is amazingly bereft of reports of the contribution of intensive care to outcome in IA. One paper did review 5.5-year admissions [70] of 25 patients with IA and respiratory failure of whom 1/3 had a normal neutrophil count 2 survived (8%); 1 who had hemoptysis from mycotic lung sequestrum and 1 given high dose ABLC with persistent neutropenia. The author's conclusion that restraint should be exercised in such patients before admitting to an ICU is a gross overstatement; considering these figures were largely pre-growth factor/lipid amphotericin B era, IA survival generally is improving, and that no reference was made to that unit's overall mortality.

Colonization, Bronchitis, Otitis and Aspergilloma

The upper respiratory tract of 10% of hospitalized patients may be colonized with *Aspergillus* species; they do not develop disease unless the patient's immune system is compromised.

Aspergillus bronchitis is a rather uncommon entity. Primary *Aspergillus* bronchitis is rare. Superinfection in patients with chronic bacterial suppurative lung disease is occasionally noted but the precise role of *Aspergillus* in this disease in still unclear. Thoracic surgical procedures may rarely be complicated by bronchial stump remnant *Aspergillus* infection. Otitis externa and media may be due to *Aspergillus* either secondarily in chronic bacterial perforating suppurative otitis media or primarily also in immunocompromised patients. In HIV patients progressive *Aspergillus* otitis involving the temporal bone may lead to facial paralysis, 6th nerve palsy and sensorineural hearing loss which may necessitate surgical intervention and systemic antifungals. Otitis externa with severe pain, facial nerve involvement and local invasion (malignant otitis externa) due to *Aspergillus* occurs mainly in immunocompromised patients and is more often due to *A. flavus* and *A. niger* than to *A. fumigatus*. Hematogenous dissemination to the lungs and other organs may complicate the outcome. Response to amphotericin B is poor – some failures appear to respond to ITRA.

Long-standing host defects of a mechanical nature, e.g. cavitary lung disease after tuberculosis or cystic fibrosis may predispose to the development of secondary aspergillomata (mycetoma). Most are asymptomatic and detected only incidentally on plain radiography. They may occasionally regress spontaneously and require no treatment. However, hemoptysis may be problematic. In this situation resective surgery is curative [72] though the underlying poor pulmonary status of many patients is an operative risk factor. In these patients embolization of the feeder arteries may be successful. Perioperative amphotericin B may reduce postoperative *Aspergillus* empyema or persistent fistula. Systemic antifungals do not penetrate into avascular cavities such as aspergillomata. The efficacy of intracavitary amphotericin B instillations has not been properly assessed [73]. Very rarely IA occurs after aspergilloma. Primary aspergilloma or mycotic lung sequestrum can arise in patients recovering from IA (fig. 3a). The necrotic fungal walls present a high (50%) risk for severe hemoptysis and death, and these patients should be managed surgically [46].

Hypersensitivity Aspergillosis

IgE type mediated responses to *Aspergillus* antigens may lead to asthma. Loeffler's syndrome (variable pulmonary infiltrates, fever, cough) is another manifestation. Management involves bronchodilators and steroids and avoiding allergen contact.

Extrinsic Allergic Alveolitis (EAA)

This occurs following repeated exposure to *Aspergillus* antigens as a result of type 3 and 4 reactions, e.g. Farmer's lung. It manifests as cough, dyspnea, fever, chills and myalgia occurring a few hours after exposure to the source. If unrecognized, severe distal airways granulomatous inflammation will lead to pulmonary fibrosis. Removal from the source, and systemic steroids are the mainstay of management.

Allergic Bronchopulmonary Aspergillosis (ABPA) and Allergic Aspergillus *Sinusitis*

Some individuals mount an extremely aggressive host damaging immunologic response to *A. fumigatus*, characterized by high eosinophilia, specific anti-*Aspergillus* IgG and IgE, upper lobe transient pulmonary infiltrates, central bronchiectasis, positive serum *Aspergillus* precipitating antibodies and type 1 skin reaction. This is ABPA [75]. Up to 10% patients with asthma have ABPA.

Fig. 5. a Skin hematogenous lesion of *Aspergillus* in a neutropenic patient. *b* Progression to severe tissue destruction and edema despite antifungals and growth factors. *c* The same patient after 14 dives in a hyperbaric oxygen chamber.

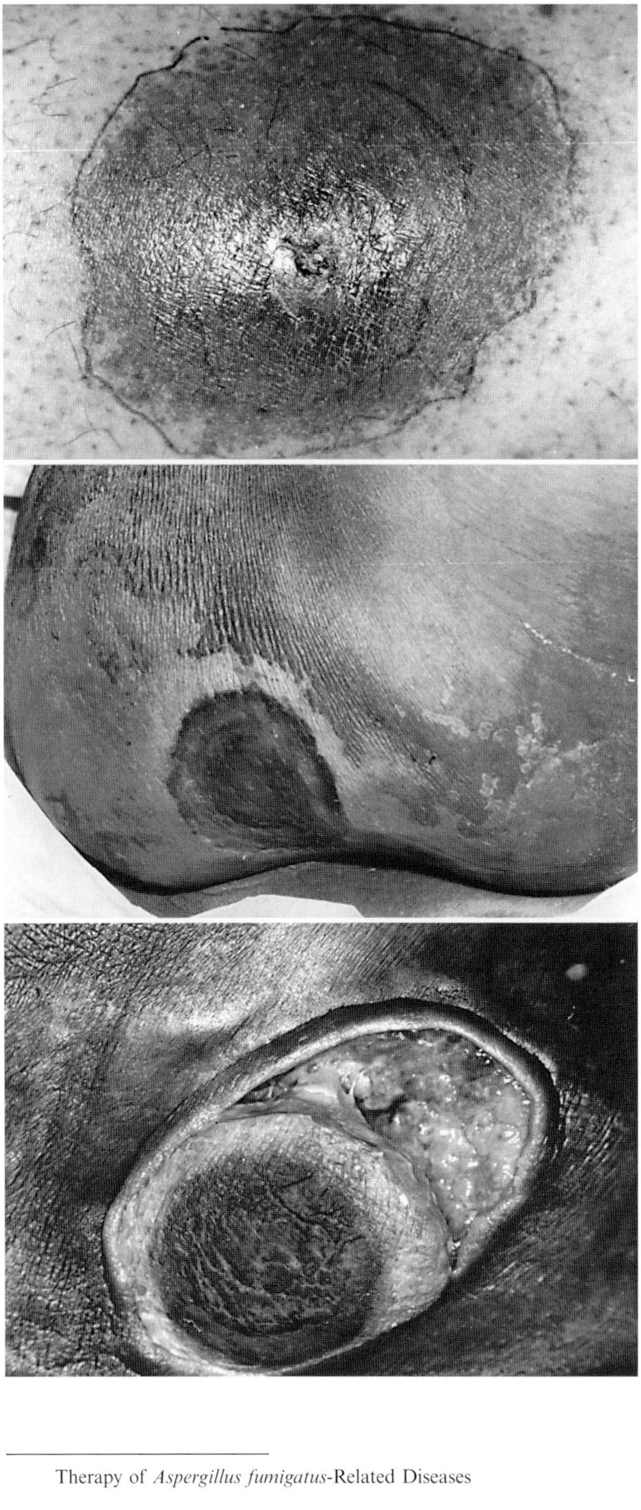

a

b

c

A recently recognized entity of allergic *Aspergillus* sinusitis is of similar pathogenesis [76]. The intense eosinophilia directly damages respiratory epithelium and lung matrix protein ultimately leading to irreversible airway damages. Treatment pivots around the use of intermittent/prolonged systemic steroids to improve respiratory symptoms, reduce pulmonary infiltrates and risk of pulmonary fibrosis. In addition, bronchodilatory therapy, postural drainage with bronchial toiletting and removal from high spore environments are important measures [75, 76]. Antifungal therapy may provide transient clinical benefit [77].

Chronic Necrotizing Pulmonary Aspergillosis

This is a slowly progressive destructive infection occurring in patients with no major underlying systemic immunosuppression but who are often debilitated, have chronic lung disease, or who are alcoholic [78]. Fungi are encased within a fibrous avascular environment, presenting difficulty for systemically administered antifungals to adequately penetrate. Clinical features of weight loss, fever, cough, tissue necrosis and predominantly upper lobe involvement may lead to an erroneous diagnosis of tuberculosis. Despite administration with systemic amphotericin B alone or in combination with other antifungals, response is poor. Surgical resection may be useful for limited disease in an attempt to prevent progression to respiratory failure and cor pulmonale [78, 79].

Future Therapeutics

Several avenues of research are posing new possibilities and challenges for the future of antifungal therapy [80]. These include first the exploitation of existing antifungal agents, e.g. voriconazole in salvage trials [81]. Some cases of intracranial IA clinically resistant to high-dose intravenous, itrathecal and intralesional CAB have responded to voriconazole [82]. A comparative randomized study of voriconazole vs. CAB is underway under the auspices of the EORTC. Other similar agents include BMS 207147, TAK 187 and UR 9825. Molecular stearic reconfiguration of existing antifungal compounds provides the potential for optimizing cost efficacy, e.g. pegylated liposomal amphotericin B is less opsonized but retains antifungal activity; dose requirements could be reduced leading to cost savings [83]. Second, clinical trials of recently identified fungal targets have been initiated. The first clinical efficacy study of the echinocandin MK991 (in candidiasis) has recently shown good results [84]. The domain 3 derived peptide XMP 391 from bactericidal permeability increasing protein has demonstrated enhanced survivorship in combination with low-dose CAB [85]. Third, preclinical exploration is underway of recently discovered antifungal compounds. These include aged cheddar cheese *Lactococcus* species [86], Cecropin lytic 37 amino acid peptide [87] – lethal to germinating *Aspergillus* via its lipopolysaccharide binding. Fourth, the delineation of 'essential' disease genes, e.g. for signal transduction, could provide the basis for development of novel drugs and vaccines [88].

Summary and Outlook

The optimal management of IA remains unclear. However, trends have emerged suggesting that outcome is improving. This has coincided with energetic attempts at making the diagnosis as early as possible and instituting treatment not only with maximally tolerated dosages of antifungals, but coupled to appropriate and timely surgery, with appropriate augmentation of host immunity. Not only are national and international guidelines needed to clarify definitions, and to guide therapy, but multicentered unified cooperative clinical studies are urgently required to logicize and optimize management.

References

1 Fridkin SC, Jarvis WR: Epidemiology of nosocomial infections. Clin Microbiol Rev 1996;9:499–511.

2 Gerson SL, Talbot GH, Hurwitz S, Lusk EJ, Strom BL, Cassileth PA: Discriminant scorecard for diagnosis of invasive pulmonary aspergillosis in patients with acute leukemia. Am J Med 1985;79:57–64.

3 Fisher BD, Armstrong D, Yu B, Gold JWM: Invasive aspergillosis: Progress in early diagnosis and treatment. Am J Med 1981;71:571–577.

4 De Repentigny I, Kilanowski E, Pednault L, Boushira M: Immunoblot analyses of the serological response to *Aspergillus fumigatus* antigens in experimental invasive aspergillosis. J Infect Dis 1991; 336:1305–1311.

5 Patterson R, Greenberger PA, Halwig JM: Allergic bronchopulmonary aspergillosis: Natural history and classification of early disease by serologic and radiographic studies. Arch Intern Med 1986; 146:916–921.

6 Rogers TR, Haynes KA, Barnes RA: Value of antigen detection in predicting invasive pulmonary aspergillosis. Lancet 1990;336:1210–1213.

7 Obayashi T: The G-test to diagnose fungal infections. Abstr O-12, Trends Invas Fung Infect 4, Barcelona, 1997.

8 Khuong A: Detection of *Aspergillus* species in bronchoalveolar lavage samples by competitive PCR. J Clin Microbiol 1995;33:1164–1168.

9 Buchheidt D, Skladny H, Baust C, Hehlmann R: Early detection of *Aspergillus* species by PCR in blood and bronchoalveolar lavage samples of immunocompromised patients. Abstr P-65, Trends Invas Fung Infect 4, Barcelona, 1997.

10 Kuhlman JE, Fishman EK, Siegelman SS: Invasive pulmonary aspergillosis in acute leukemia: Characteristic findings on CT, the CT halo sign, and the role of CT in early diagnosis. Radiology 1985;157:611–614.

11 Caillot D, Casanovas O, Bernard A, Couailler JF, Durand C, Cuisenier B, Solary E, Piard F, Pterella T, Bonnin A, Coillault G, Dumas M, Guy H: Improved management on invasive pulmonary aspergillosis in neutropenic patients using early thoracic computed tomographic scan and surgery. J Clin Oncol 1997;15:139–147.
12 Von Eiff M, Roos N, Schulten R, Hesse M, Zuhlsdorf M, van de Loo J: Pulmonary aspergillosis: Early diagnosis improves survival. Respiration 1995;62:341–347.
13 Gallis HA, Drew RH, Pickard WW: Amphotericin B: 30 years of clinical experience. Rev Infect Dis 1990;12:308–329.
14 Denning D: Therapeutic outcome in invasive aspergillosis. Clin Infect Dis 1996;23:608–615.
15 Denning DW, Stevens DA: Antifungal and surgical treatment of invasive aspergillosis: Review of 2,121 published cases. Rev Infect Dis 1990;12:1147–1201.
16 Khoo SH, Denning DW: Invasive aspergillosis in patients with AIDS. Clin Infect Dis 1994;19(suppl 1):S41–48.
17 Walsh TJ, van Cutsem J, Polak AM, Graybill JR: Immunomodulation and antifungal therapy of experimental invasive candidosis, histoplasmosis and aspergillosis: Recent advances and concepts. J Med Vet Mycol 1992;30(suppl 1):225–240.
18 Allende MA, Lee JW, Francis P, Garrett K, Dollenberg H, Berenguer J, Lyman CA, Pizzo PA, Walsh TJ: Dose-dependent antifungal activity and nephrotoxicity of amphotericin B colloidal dispersion in experimental pulmonary aspergillosis. Antimicrob Agents Chemother 1994;38:518–522.
19 Burch PA, Karp JE, Merz WG, Huhlman JE, Fishman EK: Favorable outcome of invasive aspergillosis in patients with acute leukemia. J Clin Oncol 1987;5:1985–1993.
20 Ruchlemer R, Yinnon AM, Hershko C: Changes in the natural history on invasive pulmonary aspergillosis in neutropenic leukemic patients. Israel J Med Sci 1996;32:1089–1092.
21 Brajtburg J, Bolard J: Carrier effects on biological activity of amphotericin B. Clin Microbiol Rev 1996;9:512–531.
22 Stevens DA: Overview of amphotericin B colloidal dispersion (Amphocil). J Infect 1994;28(suppl 1): 45–49.
23 Adler-Moore J: AmBisome targetting to fungal infections. Bone Marrow Transplant 1994;14 (suppl 5):S3–7.
24 Leenders ACAP, de Marie S, ten Kate MT, Bakker-Woudenberg IAJM, Verbrugh HA: Liposomal amphotericin B (AmBisome) reduces dissemination of infection as compared with amphotericin B deoxycholate (Fungizone) in a rat model of pulmonary aspergillosis. J Antimicrob Chemother 1996; 38:215–225.
25 Francis P, Lee JW, Hoffman A, Peter J, Francesconi A, Bacher J, Shelhamer J, Pizzo PA, Walsh TJ: Efficacy of unilamellar liposomal amphotericin B in treatment of pulmonary aspergillosis in persistently granulocytopenic rabbits: The potential role of bronchoalveolar D-mannitol and serum galactomannan as markers of infection. J Infect Dis 1994;169:356–368.
26 Hiemenz JW, Walsh TJ: Lipid formulations of amphotericin B: Recent progress and future directions. Clin Infect Dis 1996;(suppl 2):S133–144.
27 Ng TTC, Denning DWW: Liposomal amphotericin B (AmBisome) therapy in invasive fungal infections. Evaluation of United Kingdom compassionate use data. Arch Intern Med 1995;155: 1093–1098.
28 Oppenheim BA, Herbrecht R, Kusne S: The safety and efficacy of amphotericin B colloidal dispersion in the treatment of invasive mycosis. Clin Infect Dis 1995;21:1145–1153.
29 Mills W, Chopra R, Linch DC, Goldstone AH: Liposomal amphotericin B in the treatment of fungal infections in neutropenic patients: A single-centre experience of 133 episodes in 116 patients. Br J Haematol 1994;86:754–760.
30 Oracova E, Mistrik M, Sakalova A: Amphotericin B lipid complex to treat invasive fungal infections in cancer patients: Report of efficacy and safety in 20 patients. Chemotherapy 1995;41:473–476.
31 White MH, Anaissie EJ, Kusne S, Wingard JR, Hiemenz JW, Cantor A, Gurwith M, Du Mond C, Mamelok RD, Bowden RA: Amphotericin B colloidal dispersion vs amphotericin B as therapy for invasive aspergillosis. Clin Infect Dis 1997;24:635–642.
32 Akhan SA, Marinus A, De Pauw BE: An analysis of definitions on invasive fungal infections in immunocompromised patients. Abstr P-1, Trends Invas Fung Infect 4, Barcelona, 1997.

33 Ellis M, Spence D, De Pau B, Meunier F, Marinus A, Collette L, Sylvester R, Meis J, Boogaerts M, Selleslag D, Krcmery V, von Sinner W, Macdonald P, Doyen C, Vandercam B: Invasive aspergillosis in neutropenic patients: A randomised international multi-centre treatment study comparing 1 mg/kg per day versus 4 mg/kg per day of liposomal amphotericin B (AmBisome) (EORTC No 19923): Abstr LM39. 36th Intersci Conf Antimicrob Agents Chemother (ICAAC), New Orleans, 1996.

34 Prentice HG, Hann IM, Herbrecht R, Aoun M, Kvaloy S, Catovsky D, Pinkerton CR, Schey SA, Jacobs F, Oakhill A, Stevens RF, Darbyshire PJ, Gibson BES: A randomised comparison of liposomal versus conventional amphotericin B for the treatment of pyrexia of unknown origin in neutropenic patients. Br J Haematol 1997;98:711–718.

35 Hann IM, Stevens RF, Pinkerton CR, Darbyshire PJ, Gibson BES, Oakhill A: Safety and efficacy of two dose regimes of AmBisome versus amphotericin B as empiric antifungal treatment in neutropenic paediatric patients. Abstr, 2nd International Symp Febrile Neutropenia, Brussels, 1995.

36 Walsh T, Bodensteiner D, Hiemenz J, Thaler S, Greenberg R, Arndt C, Holcenberg J, Schwartz C, Pappas P, Dummer S, Marsh R, Schuster M, Seibel N: The mycoses study group (MSG), the empirical antifungal study group. Abstr LM-90, 37th Intersci Conf Antimicrob Agents Chemother (ICAAC), Toronto, 1997.

37 Leenders ACAP, Daenen S, Jansen RLH, Hop WC, Lowenberg B, Cornelissen J, Herbrecht R, De Marie S: Liposomal amphotericin B (AmBisome) compared with amphotericin B in the treatment on neutropenia-associated fungal infections. Abstr P-6, Trends Invas Fung Infect 4, Barcelona, 1997.

38 Van't Wout JW: Itraconazole in neutropenic patients. Chemotherapy 1992;(suppl 1):23–26.

39 Denning DW, Lee JY, Hostetler JS, Pappas P, Kauffman CA, Dewsnup DH, Galgiani JN, Graybill JR, Sugar AM, Catanzaro A: NIAID mycoses study group multicenter trial of oral itraconazole therapy for invasive aspergillosis. Am J Med 1994;97:135–144.

40 Sanchez C, Mauri E, Dalmau D, Quintana S, Aparicio A, Garau J: Treatment of cerebral aspergillosis with itraconazole: Do high doses improve the prognosis? Clin Infect Dis 1995;21:1485–1487.

41 Verweij PE, Donnelly JP, Kullberg BJ, Meis JF, de Pauw BE: Amphotericin B versus amphotericin B plus 5-flucytosine: Poor results in the treatment of proven systemic mycoses in neutropenic patients. Infection 1994;22:81–85.

42 Maesaki S, Kohno S, Kaku M, Koga H, Hara K: Effects of antifungal agent combinations administered simultaneously and sequentially against *Aspergillus fumigatus*. Antimicrob Agents Chemother 1994;38:2843–2845.

43 Warnock DW: Antifungal drug susceptibility testing and resistance in *Aspergillus*. Abstr O-24, Trends Invas Fung Infect 4, Barcelona, 1997.

44 Ellis M: Fungal endocarditis. J Infect 1997;35:99–103.

45 Wong K, Waters CM, Walesby RK: Surgical management of invasive pulmonary aspergillosis in immunocompromised patients. Eur J Cardio-Thorac Surg 1992;6:138–143.

46 Robinson LA, Reed EC, Galbraith TA, Alonso A, Moulton AL, Fleming WH: Pulmonary resection for invasive *Aspergillus* infections in immunocompromised patients. J Thorac Cardiovasc Surg 1995; 109:1182–1197.

47 Buescher TM, Moritz DM, Killyon GW: Resection of chest wall and central veins for invasive cutaneous *Aspergillus* infection in an immunocompromised patient. Chest 1994;105:1283–1285.

48 Coleman JM, Hogg GG, Rosenfeld JV, Waters KD: Invasive central nervous system aspergillosis: Cure with liposomal amphotericin BV, itraconazole and radical surgery: Case report and review of the literature. Neurosurgery 1995;36:858–863.

49 Karp JE, Burch PA, Merz WG: An approach to intensive antileukemia therapy in patients with previous invasive aspergillosis. Am J Med 1988;85:203–206.

50 Saeed SR, Brookes GB: Aspergillosis of the paranasal sinuses. Rhinology 1995;33:46–51.

51 Min YG, Kim HS, Lee KS, Kang MK, Han MH: *Aspergillus* sinusitis: Clinical aspects and treatment outcomes. Otolaryngol Head Neck Surg 1996;115:49–52.

52 Rowe-Jones JM, Freedman AR: Adjuvant itraconazole in the treatment of destructive sphenoid aspergillosis. Rhinology 1994;32:203–207.

53 Ur-Rahman N, Jamjoom A, Al-Hedaithy, Jamjoom ZAB, Al-Sohaibani MO, Aziz SA: Cranial and intracranial aspergillosis of sino-nasal origin. Acta Neurochir (Wien) 1996;138:944–950.

54 Gumaa SA, Mahgoub ES, Hay RJ: Post-operative responses of paranasal *Aspergillus* granuloma to itraconazole. Trans R Soc Trop Med Hyg 1992;86:93–94.
55 De Carpentier JP, Ramamurthy L, Denning DW, Taylor PH: An algorithmic approach to *Aspergillus* sinusitis. J Laryngol Otol 1994;108:314–318.
56 Landoy Z, Rotstein C, Shedd D: Aspergillosis of the nose and paranasal sinuses in neutropenic patients at an oncology centre. Head Neck Surg 1985;8:83–90.
57 Kavanagh KT, Hughes WT, Parham DM, Chanin LR: Fungal sinusitis in immunocompromised children with neoplasms. Ann Otol Rhinol Laryngol 1991;100:331–336.
58 Wiatrak BJ, Willging P, Myer CM III, Cotton RT: Functional endoscopic sinus surgery in the immunocompromised child. Otolaryngol Head Neck Surg 1991;105:818–825.
59 The W, Matti BS, Marisiddaiah H, Minamoto GY: *Aspergillus* sinusitis in patients with AIDS: Report of three cases and review. Clin Infect Dis 1995;21:529–535.
60 Nicholson AG, Sim KM, Keogh BF, Corrin B: Pseudomembranous necrotising bronchial aspergillosis complicating chronic airways limitation. Thorax 1995;50:807–808.
61 Sugata T, Myoken Y, Kyo T, Fujhara M: Invasive oral aspergillosis in immunocompromised patients with leukemia. J Oral Maxillofac Surg 1994;52:382–386.
62 Isaac M: Cutaneous aspergillosis. Dermatol Clin 1996;14:137–140.
63 Goodman ML, Coffey RJ: Stereotactic drainage of *Aspergillus* brain abscess with long-term survival: Case report and review. Neurosurgery 1989;24:96–99.
64 Camarata PJ, Dunn DL, Farney AC, Parker RG, Seljeskog EL: Continual intracavitary administration of amphotericin B as an adjunct in the treatment of *Aspergillus* brain abscess: Case report and review of the literature. Neurosurgery 1992;31:575–579.
65 Chanock SJ, Gorlin JB: Granulocyte transfusions: Time for a second look. Infect Dis Clin N Am 1996;10:327–343.
66 Stevens DA: Immunomodulation as a form of antifungal therapy. Abstr S-109, 37th Intersci Conf Antimicrob Agents Chemother, Toronto, 1997.
67 Bodey GP, Anaissie E, Gutterman J, Vadhan-Raj S: Role of granulocyte-macrophage colony-stimulating factor as adjuvant therapy for fungal infection in patients with cancer. Clin Infect Dis 1993;17:705–707.
68 Cenciu E, Perito S, Enssle KH, Mosci P, Latge JP, Romani L, Bistoni F: Th1 and Th2 cytokines in mice with invasive aspergillosis. Infect Immun 1997;65:564–570.
69 Price JC, Stevens DL: Hyperbaric oxygen in the treatment of rhinocerebral mucormycosis. Laryngoscope 1980;90:737–747.
70 Janssen JJWM, Strack van Schijndel RJM, van der Poest Clement EH, Ossenkoppele GJ, Thijs LG, Huigens PC: Outcome of ICU treatment in invasive aspergillosis. Intens Care Med 1996;22:1315–1322.
71 Harley WB, Dummer JS, Anderson TL, Goodman S: Malignant external otitis due to *Aspergillus flavus* with fulminant dissemination to the lungs. Clin Infect Dis 1995;20:1052–1054.
72 Daly RC, Pairolero PC, Piehler JM: Pulmonary aspergilloma: Results of surgical treatment. J Thorac Cardiovasc Surg 1986;92:981–988.
73 Ryan PJ, Stableforth DE, Reynolds J, Muhdi KM: Treatment of pulmonary aspergilloma in cystic fibrosis by percutaneous instillation of amphotericin B via indwelling catheter. Thorax 1995;50:809–810.
74 Fuller CJ: Farmer's lung. Dis Chest 1962;42:176–180.
75 Greenberger PA, Patterson R: Diagnosis and management of allergic bronchopulmonary aspergillosis. Ann Allkerg 1986;56:444–448.
76 Waxman JE, Spector JG, Sale SR, Katzenstein A: Allergic *Aspergillus* sinusitis: Concepts in diagnosis and treatment of a new clinical entity. Laryngoscope 1987;97:261–266.
77 Fournier EC: Trial of ketoconazole in allergic bronchopulmonary aspergillosis. Thorax 1987;42: 831–835.
78 Binder RE, Faling J, Pugatch RD, Mahassen C, Snider SL: Chronic necrotising pulmonary aspergillosis: A discrete clinical entity. Medicine 1982;61:109–124.
79 Ellis ME, Dossing M, Al-Hokail A, Qadri SH, Hainau B, Burnie J: Progressive chronic pulmonary aspergillosis: A diagnostic and therapeutic challenge. J R Soc Med 1991;763–744.
80 Georgopaadakou NH, Walsh TJ: Antifungal agents: Chemotherapeutic targets and immunologic strategies. Antimicrob Agents Chemother 1996;40:279–291.

81 del Favero A, Denning DW, Gluckman E: UK-109,496 a novel wide spectrum triazole derivative for the treatment of fungal infections: Clinical efficacy in acute invasive aspergillosis. Abstr F-80, 35th Intersc Conf Antimicrob Agents Chemother, San Francisco, 1995.
82 Scwartz S, Milatovic D, Thiel E: Successful treatment of cerebral aspergillosis with a novel triazole (voriconazole) in a patient with acute leukaemia. Br J Haematol 1997;97:663–665.
83 Van Etten EWM, Kate MTT, Bakker-Woundfenberg IAJM: Efficacy of a new type of liposomal amphotericin B in severe invasive pulmonary aspergillosis in leukopenic rats. Abstr B-15, 37th Intersci Conf Antimicrob Agents Chemother, Toronto, 1997.
84 Sable CA, Villaneuva A, Arathon E, Gotuzzo E, Turcato G, Uip D, Noriega L, Rivera C, Rojas E, Taylor V, Berman R, Calandra GB, Chodakevitz J: A randomised double blind multicenter trial of MK-991 (L-743,872) vs amphotericin B (AMB) in the treatment of *Candida* esophagitis in adults. Abstr LB-33, 37th Intersci Conf Antimicrob Agents Chemother, Toronto, 1997.
85 Ammons WS, Aardalen K, Froebel S, Little R: Efficacy of a domain III-derived peptide from bactericidal/permeability-increasing protein (BPI) in murine disseminated aspergillosis. Abstr B-16, 37th Intersci Conf Antimicrob Agents Chemother, Toronto, 1997.
86 Roy U, Batish VK, Grover S, Neelakantan S: Production of antifungal substance by *Lactococcus lactis subspp. lactis* CHD-28. 3. Int J Food Microbiol 1996;32:27–34.
87 DeLucca AJ, Bland JM, Jacks TJ, Grimm C, Cleveland TE, Walsh TJ: Fungicidal activity of cecropin A. Antimicrob Agents Chemother 1997;41:481–483.
88 Perfect JR: *Cryptococcus neoformans*: A molecular model for antifungal target development. Abstr S-107, 37th Intersci Conf Antimicrob Agents Chemother, Toronto, 1997.

Dr. Michael Ellis, Division of Infectious Diseases, Department of Medicine,
Tawam Hospital, Al-Ain, Abu Dhabi (United Arab Emirates)
Tel. +9731 677 444

Brakhage AA, Jahn B, Schmidt A (eds): *Aspergillus fumigatus*.
Contrib Microbiol. Basel, Karger, 1999, vol 2, pp 130–138

Animal Models of *A. fumigatus* Infections

Kazutoshi Shibuya[a], *Shiro Naoe*[a], *Hideyo Yamaguchi*[b]

[a] Department of Pathology, Toho University Ohashi Hospital, and
[b] Department of Microbiology, Teikyo University School of Medicine, Tokyo, Japan

Aspergillosis can be defined as damage to human or animal tissue caused by several species of *Aspergillus*. Damage may be caused by invasion of tissues, colonization of cavities, or allergic mechanisms, resulting in diseases which vary in severity and clinical course depending on the organs affected and the host. Aspergilli are abundant in the environment. The genus *Aspergillus* includes more than 200 recognized species, all of which are distributed throughout the world. Their conidia are produced in great abundance and are readily disseminated into the air by wind currents. Although frequently inhaled by man, most cases of human aspergilloses are thought to be caused by *A. fumigatus* [1, 2].

Due to the respiratory tract usually being a portal of infection the lung is the most common site affected. Individuals with apparently normal defense mechanisms rarely get infected with *A. fumigatus*. Systemic aspergillosis followed by pulmonary infection occurs mostly in immunocompromised individuals such as patients with chronic granulomatous diseases, leukemia and other hematological neoplasia, bone marrow or other transplant patients, and in individuals receiving corticosteroids, immunosuppressive drugs, and/or broad-spectrum antibiotics [2–4]. It has been concluded from clinical observations that neutropenia and high doses of corticosteroids are the two major risk factors for invasive aspergillosis [4, 5] and that in combination they may act synergistically to break down natural resistance [4]. This form of invasive aspergillosis is mainly implicated by *A. fumigatus*, with *A. flavus* as the second most important pathogen [2].

Pathology of Pulmonary Aspergillosis

Pulmonary aspergillosis has been classified into three forms: noninvasive, invasive, and allergic. Preexisting or concurrent disease resulting in anatomical alterations of the lung is a major factor predisposing a host to *A. fumigatus* colonization which usually results in the noninvasive form of pulmonary aspergillosis. A tuberculosis cavity, lung abscess, cavitating infarcts, lung cysts, bronchiectasis, and sarcoidosis all predispose the patient to *Aspergillus* colonization [6]. By the time the condition is diagnosable, the fungus has usually grown to form a compact mass of mycelium termed 'fungus ball' or 'aspergilloma'. In most cases of noncompromised host, the fungus does not invade the surrounding tissue or spread to other parts of the body. However, if an aspergillous infection occurred in an immunocompromised patient, both with and without preexisting lung disease, parenchymal pneumonia, usually indicated as bronchopneumonia, develops in the lung which may result in a systemic disease. In recent years, clinical and pathological evidence of invasive pulmonary aspergillosis (IPA) has been recognized as one of the important opportunistic fungal infections. The disease raises a serious problem among some specific immunocompromised populations. The variety of clinical and radiological manifestations of IPA may depend on the immunological situation of the host [3]. Namely, a fulminant form of IPA can be found when the infection develops in patients with prolonged agranulocytosis [4] (fig. 1). In addition, chronic necrotizing pulmonary aspergillosis [7] and some other clinical forms of pulmonary aspergillosis [3] have been reported which may occur in mildly immunocompromised patients, e.g. those with diabetes mellitus and those receiving low doses of steroids [7].

Allergic bronchopulmonary aspergillosis (ABPA) is known as one of the major forms of pulmonary aspergillosis. This allergic disease is being recognized with increasing frequency [8–10]. It occurs in previously sensitized individuals who are exposed to conidia of *Aspergillus* species, especially *A. fumigatus*, as the most usual etiological agent [10]. In ABPA, febrile episodes occur in association with transient lung infiltrates recurring at multiple sites. This syndrome is associated with pulmonary consolidation and eosinophilia.

Animal Models of *A. fumigatus* Infection

A major aim of an experimental study of aspergillosis with animal models is to gain insight into the pathogenesis of human aspergillosis. However, experimental models have also been used for the evaluation of drug efficacy, elucida-

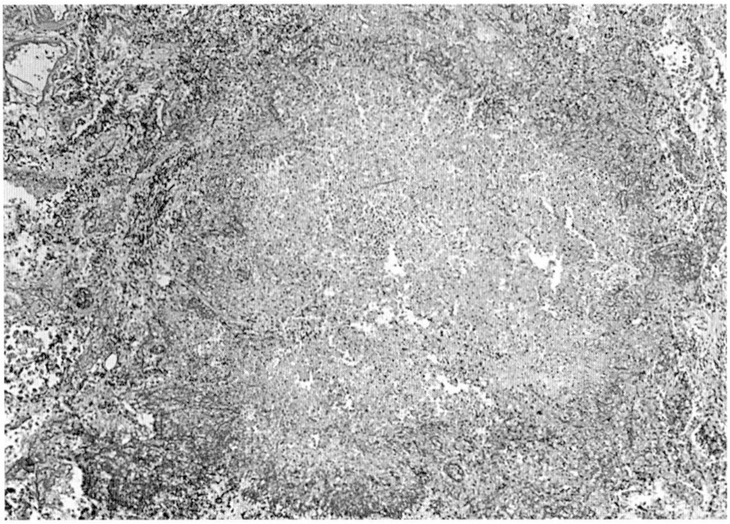

Fig. 1. Micrograph showing a lesion of invasive pulmonary aspergillosis in the human lung. The center of the lesion appears necrotic with prominent neutrophilic infiltrate. Hyphae of aspergilli align in a radial pattern in the periphery of this lesion. PAS-Elastica. × 180.

tion of virulent factors of aspergilli, assessment of new diagnostic procedures involving serological, molecular biological and histopathological examination, and immunological analysis of the infection. A large number of studies have been conducted, using a variety of experimental models of the disease. Fatal experimental infections leading to multiple organ involvement have been most frequently produced in mice with an intravenous injection of conidia. This type of murine model has been demonstrated to be effective for evaluating antifungal agents; although less frequently, rabbits and various other species of animals, including rats, monkeys, sheep, ducks and chicks have also been employed as models for this type of experimental aspergillosis [11].

Since mice, rabbits, and other laboratory animals normally show a high degree of resistance towards *Aspergillus* infections, their host defense can be lowered by the administration of immunosuppressive agents or some other way in order to ensure establishment of persistent and/or invasive infections in these animals. Animals with predisposition to experimental *Aspergillus* infection and, thus, often used to produce animal models or systemic aspergillosis are those with alloxin-induced diabetes [12], those fed deficient diets [13], as well as those treated with several immunosuppressive drugs, such as corticosteroids, cyclosporin A and cyclophosphamide [14]. Species of animals, routes of infection and immunosuppressive agents must be adequately chosen.

The animal models of aspergillous infection have been mainly produced in the three different species of rodents: mice, rabbits and rats.

Murine Models of Aspergillosis

The most frequently used animal model of aspergillosis is that produced in the mouse which has various merits for experimental studies. In addition to the convenience of laboratory handling, various inbred strains of which immunological and genetical defects have been clearly defined are easily obtainable. Accordingly, inbred mice can be the most suitable animal to analyze defense mechanisms and cytokine cascades of the host against *Aspergillus*-associated infections. Mice lethally infected by intravenous inoculation with a conidial suspension have been commonly used to evaluate the efficacy of novel antifungal drugs on the basis of mortality and organ cultures, as well as to test the virulence of different strains of aspergilli [11]. More recently, however, such intravenous models have been used less frequently [15]. Currently, the most frequently chosen route of infection is via the respiratory tract because it reflects the natural route of aspergillous infection in humans [5, 16]. Most of these murine models, usually reported as a model of IPA, are developed in animals pretreated with immunosuppressive and/or myelotoxic agents to induce granulocytopenia. Mostly, mice are inoculated with a conidial suspension intranasally, but airborne models in which mice are infected by the aspiration of aerosols containing *Aspergillus* conidia have also been reported [16]. If the usage of such experimental infections were limited to an induction of primary lesions in the lungs, these intranasal models might be satisfactory.

Recently, several *A. fumigatus* strains, genetically different from their parental wild-type strains in term of genes encoding putative virulence factors, have been developed and produced by gene disruption techniques and applied to produce experimental murine models of pulmonary infection in which a conidial suspension was inoculated intranasally [5, 16–19]. However, despite using these molecular biological approaches, the most important virulent determinant of aspergilli in pulmonary lesions has still not been made clear. The not frequently reported histopathological findings of pulmonary lesions in mice appear to be somewhat different from that seen in corresponding human disease, namely IPS. As previously mentioned, the essential histopathological feature of IPA in humans can be defined as a bronchopneumonia which results from both a proliferation of aspergilli and an exudative response in the alveoli. However, in our experience, development of most aspergillous lesions in the lung of infected mice initiated from conidia which settled on the bronchial mucosa. This histological difference in lesions between mice and humans may be due to the anatomical simplicity of the bronchial tree in mice and the hydrophobicity of *Aspergillus* conidia. Accordingly, when animal models are

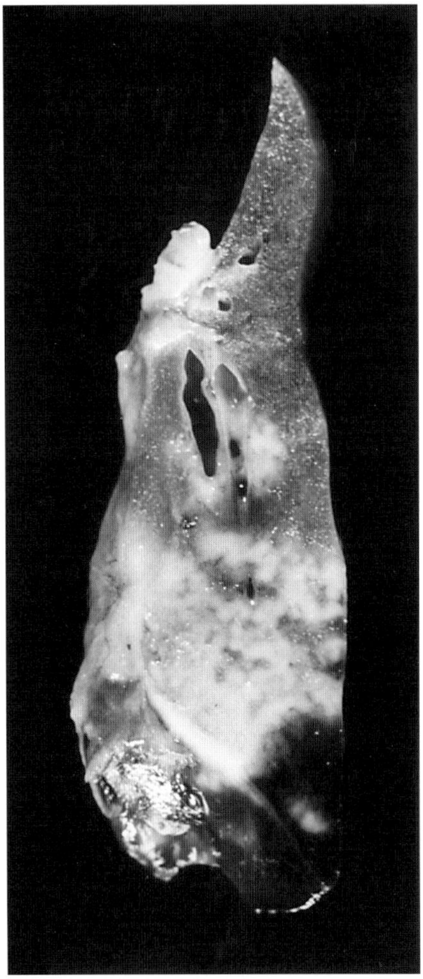

Fig. 2. Gross feature of invasive pulmonary aspergillosis in rats achieved by intratracheal injection of a suspension of agarose beads containing *A. fumigatus* conidia. The distal segment of the upper of the left lung, corresponding to the lingual segment of the human lung, is replaced by a poorly defined solid lesion with a peripheral hemorrhage.

required where the histopathological features as well as the pathophysiological states more closely resemble those of human IPA, larger animals with more complex bronchial trees should be chosen.

The intraperitoneal route of inoculation has been used to induce fatal infection in mice, particularly those treated with immunosuppressive agents. However, recently this route of infection has often been used as the primer followed by respiratory tract exposure of conidia in order to elucidate an immunological response [20] or eosinophilopoiesis in bone marrow [21]; a procedure which is closely related to the pathogenesis of ABPA.

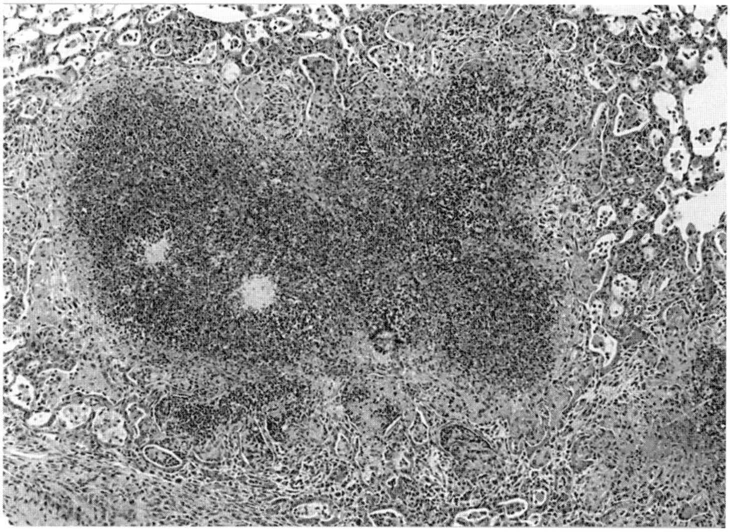

Fig. 3. Micrograph showing the lesion of experimental invasive pulmonary aspergillosis in an infected rat. The histological feature of this lesion is characterized by a necrosis containing numerous neutrophils. Hyphae of aspergilli are seen in the periphery of this lesion. The necrosis is encompassed with a response of histiocytes. PAS-Elastic. × 180.

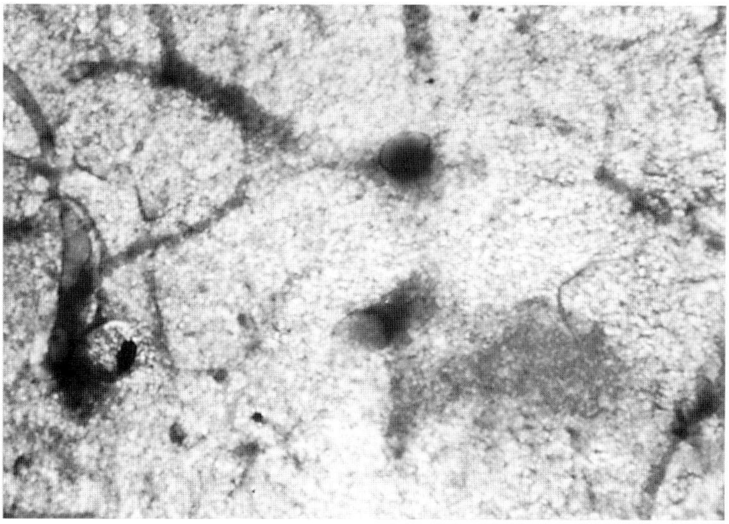

Fig. 4. Photograph of a thick section of the lung of an infected rat, showing a continuous elongation of fungal emboli in both pulmonary veins and arteries. Thick section cleared with xylene. × 18.

Rabbit Models

Next to the mouse, the rabbit has been used most frequently to generate a model of *Aspergillus* infection. Recently, rabbits have commonly been used to produce a primary pulmonary infection to elucidate the pathogenesis of IPA [14], to evaluate the clinical usefulness of newly developed antifungal agents [22, 23], and to establish and test new serodiagnostic procedures [24]. The rabbit may be large enough to accommodate a laryngoscope to inoculate a conidial suspension by inserting a catheter [14]. This may be an important merit because a short cervical incision, which is usually performed to expose the trachea of the animals especially those treated with immunosuppressive drugs, increases the risk of bacterial infection. Furthermore, rabbit models are advantageous over murine or rat models in that a rabbit is much larger than the other species of laboratory animals. This makes it possible to perform serial sampling of serum from an infected individual, detailed histopathological observation of pulmonary lesions [14], and cytological examination of bronchoalveolar lavage specimens [25]. In this sense, rabbits infected by respiratory tract injection of conidial suspension are considered the most suitable model of IPA. Further, rabbits infected by intravenous injection of a conidial suspension also provide experimental models of specific forms of systemic aspergillosis, such as endophthalmitis [26] and endocarditis [11].

Rat Models

The rat is another species of laboratory animal which is often used for experimental investigations of aspergillosis because of its suitable size to produce primary pulmonary aspergillosis infection [13, 27, 28]. We recently reported a new rat model of IPA which was produced by intratracheal injection of agarose beads containing *A. fumigatus* conidia which enabled accurate delivery to the alveoli [29] (fig. 2). The histopathological findings of the pulmonary lesions obtained with this rat model are closely similar to those for IPA in humans (fig. 3). In addition, the development of fungal emboli, which is also known to be a characteristic of the pathology of IPA, is demonstrated (fig. 4).

The recent conditions about animal models of aspergillus infection, namely focusing on the models in mice, rabbits, and rats, are discussed in this chapter.

References

1 Raper KB, Fennel DI: The genus *Aspergillus*. Baltimore, Williams & Wilkins, 1965.
2 Kwon-Chung KJ, Bennett JE: Medical mycology. Philadelphia, Lea & Febiger, 1992.
3 Gefter WB: The spectrum of pulmonary aspergillosis. J Thorac Imaging 1992;7:56–74.
4 Guiot HFL, Fibbe WE, van't Wout JW: Risk factors for fungal infection in patients with malignant hematologic disorders: Implications for empirical therapy and prophylaxis. Clin Infect Dis 1994; 18:525–532.
5 Smith JM, Tang CM, Van Noorden S, Holden DW: Virulence of *Aspergillus fumigatus* double mutants lacking restrictocin and an alkaline protease in a low-dose model of invasive pulmonary aspergillosis. Infect Immun 1994;62:5247–5254.
6 Fraser RS: Pulmonary aspergillosis: Pathologic and pathogenic features. Pathol Ann 1993;28:231–277.
7 Binder RE, Faling LJ, Pugatch RD, Mahasaen C, Snider GL: Chronic necrotizing pulmonary aspergillosis: A discrete clinical entity. Medicine 1982;61:109–124.
8 Kurup VP, Choi H, Murali PS, Coffman RL: IgE and eosinophil regulation in a murine model of allergic aspergillosis. J Leukoc Biol 1994;56:593–598.
9 Kurup VP, Seymour BW, Choi H, Coffman RL: Particulate *Aspergillus fumigatus* antigens elicit a TH2 response in BALB/c mice. J Allergy Clin Immunol 1994;93:1013–1020.
10 Murali PS, Bamrah BS, Choi H, Fink JN, Kurup VP: Hyperimmune serum modulates allergic response to spores in a murine model of allergic aspergillosis. J Leukoc Biol 1994;55: 29–34.
11 Miyaji M, Yamaguchi H: Animal Models in Medical Mycology. Boca Raton, CRC Press, 1987.
12 Ford S, Friedman L: Experimental study of the pathogenicity of *Aspergilli* for mice. J Bacteriol 1967;94:928–933.
13 Niki Y, Bernard EM, Edwards FF, Schmitt HJ, Yu B, Armstrong D: Model of recurrent pulmonary aspergillosis in rats. J Clin Microbiol 1991;29:1317–1322.
14 Berenguer J, Allende MC, Lee W, Garrett K, Lyman C, Ali NM, Bacher J, Pizzo PA, Walsh TJ: Pathogenesis of pulmonary aspergillosis: Granulocytopenia versus cyclosporin and methylpredonisoloe-induced immunosuppression. Am J Respir Crit Care Med 1995;152:1079–1086.
15 Jensen HE, Aalbaek B, Hau J, Latge JP: Detection of galactomannan and complement activation in the pregnant mouse during experimental systemic aspergillosis. APMIS 1996;104:926–932.
16 Tang CM, Cohen J, Krausz T, Van Noorden S, Holden DW: The alkaline protease of *Aspergillus fumigatus* is not a virulence determinant in two murine models of invasive pulmonary aspergillosis. Infect Immun 1993;61:1650–1656.
17 Mellado E, Specht CA, Robbins PW, Holden DW: Cloning and characterization of chsD, a chitin synthase-like gene of *Aspergillus fumigatus*. FEMS Microbiol Lett 1996;143:69–76.
18 Mondon P, De Champs C, Donadille A, Ambroise-Thomas P, Grillot R: Variation in virulence of *Aspergillus fumigatus* strains in a murine model of invasive pulmonary aspergillosis. J Med Microbiol 1996;45:186–191.
19 Frosco MB, Chase T Jr, Macmillan JD: The effect of elastase-specific monoclonal and polyclonal antibodies on the virulence of *Aspergillus fumigatus* in immunocompromised mice. Mycopathologia 1994;125:65–76.
20 Grunig G, Corry DB, Leach MW, Seymour BW, Kurup VP, Rennick DM: Interleukin-10 is a natural suppressor of cytokine production and inflammation in a murine model of allergic bronchopulmonary aspergillosis. J Exp Med 1997;185:1089–1099.
21 Murali PS, Kurup VP, Guo J, Fink JN: Development of bone marrow eosinophilia in mice induced by *Aspergillus fumigatus* antigens. Clin Immunol Immunopathol 1997;84:216–220.
22 George D, Miniter P, Andriole VT: Efficacy of UK-109496, a new azole antifungal agent, in an experimental model of invasive aspergillosis. Antimicrob Agents Chemother 1996;40:86–91.
23 Berenguer J, Ali NM, Allende MC, Lee J, Garrett K, Battaglia S, Piscitelli SC, Rinaldi MG, Pizzo PA, Walsh TJ: Itraconazole for experimental pulmonary aspergillosis: Comparison with amphotericin B, interaction with cyclosporin A, and correlation between therapeutic response and itraconazole concentrations in plasma. Antimicrob Agents Chemother 1994;38:1303–1308.

24 de Repentigny L, Kilanowski E, Pedneault L, Boushira M: Immunoblot analyses of the serologic response to *Aspergillus fumigatus* antigens in experimental invasive aspergillosis. J Infect Dis 1991; 163:1305–1311.
25 Chilvers ER, Spreadbury CL, Cohen J: Bronchoalveolar lavage in an immunosuppressed rabbit model of invasive pulmonary aspergillosis. Mycopathologia 1989;108:163–171.
26 McGuire TW, Bullock JD, Bullock JD Jr, Elder BL, Funkhouser JW: Fungal endopthalmitis: An experimental study with a review of 17 human ocular cases. Arch Ophthalmol 1991;109:1289–1296.
27 Cicogna CE, White MH, Bernard EM, Ishimura T, Sun M, Tong WP, Armstrong D: Efficacy of prophylactic aerosol amphotericin B lipid complex in a rat model of pulmonary aspergillosis. Antimicrob Agents Chemother 1997;41:259–261.
28 Kurtz MB, Bernard EM, Edwards FF, Marrinan JA, Dropinski J, Douglas CM, Armstrong D: Aerosol and parenteral pneumocandins are effective in a rat model of pulmonary aspergillosis. Antimicrob Agents Chemother 1995;39:1784–1789.
29 Shibuya K, Ando T, Wakayama M, Takaoka M, Uchida K, Naoe S: Pathological spectrum of invasive pulmonary aspergillosis: Study of pulmonary lesions of 54 autopsies and the relationship between neutrophilic response and histologic features of lesions in experimental aspergillosis. Jpn J Med Mycol 1997;38:175–181.

Kazutoshi Shibuya, Department of Pathology, Toho University Ohashi Hospital,
2–17–6 Ohashi, Meguro-ku, Tokyo 153-8515 (Japan)
Tel. +81 3 3468 1251, Fax +81 3 3468 1283, E-Mail kaz@med.toho-u.ac.jp

Brakhage AA, Jahn B, Schmidt A (eds): *Aspergillus fumigatus*.
Contrib Microbiol. Basel, Karger, 1999, vol 2, pp 139–148

Problems of Antifungal in vitro Testing in *Aspergillus fumigatus*

A. Espinel-Ingroff

Medical College of Virginia/Virginia Commonwealth University,
Richmond, Va., USA

The genus *Aspergillus*, including the species *Aspergillus fumigatus* and *Aspergillus flavus*, was first described in 1729 by the Florentine botanist, Micheli, and the role of *Aspergillus* spp. as etiologic agents of disease in human pulmonary disease was first reported by the German physician, Virchow in 1856 [1]. During the 1950s and 1960s, *Aspergillus* endocarditis and other invasive aspergillosis were associated with the constant administration of penicillin, other antibiotics and steroids. An increased awareness of the prevalence of mycoses, including aspergillosis, in cancer and leukemia patients became evident in the early 1960s; aspergillosis was the cause of death in 37% of these patients [1]. During the last 20 years, the incidence and prevalence of these infections has increased among the larger populations of bone marrow transplant, leukemia and AIDS patients as well as noncompromised individuals. *A. fumigatus* is responsible for the majority (85–90%) of the different clinical manifestations of *Aspergillus* infections, which are difficult to diagnose and treat; a high number of patients die despite antifungal therapy [2]. Only five antifungal agents are presently available or licensed for the systemic treatment of *Aspergillus* infections, amphotericin B and its lipid formulations abelcet, amphocil and ambisome and the triazole, itraconazole. However, new triazoles and chitin and glucan synthesis inhibitor agents are at different stages of clinical evaluation. These new agents are targeted to fungi that are resistant to established treatment.

With the use of both established and investigational agents has come the recognition of resistance in selected isolates and clinical laboratories have a greater role in the selection and monitoring of antifungal therapy. In vitro susceptibility testing provides a quantitative measurement or the minimum inhibitory concentration (MIC). Traditionally, MIC values were considered the ultimate predictors of clinical response to antimicrobial therapy. This concept should be revised due to the following factors. Clinical failure to antifungal therapy may be the result of microbiologic resistance or associated with drug pharmacokinetics including absorption, dosage, penetration, drug-drug interactions and patient compliance; factors related to the host immune response and/or status of an underlying disease; proper patient management; site and severity of infection; and factors related to the virulence of the infecting strain and its interactions with both the host and the therapeutic agent [3]. Therefore, since the MIC is not a chemical measurement and many factors can influence response to therapy, a low MIC does not necessarily predict clinical success. However, it is expected that in vitro resistance may be able to identify among a population of susceptible strains those strains which are less likely to respond to a specific antifungal therapy. With these concepts in mind, the evolution and problems of in vitro testing for *A. fumigatus* since 1956 were evaluated.

Antifungal Agents

Non-water-soluble agents like intraconazole and amphotericin B may be solubilized in a variety of solvents. The actual amount to be weighed must be adjusted to the specific biological activity of each standard of established and investigational agents. The non-water-soluble agents will not remain completely solubilized in the aqueous media used for susceptibility testing. To avoid this problem and obtain an equal concentration of solvent in each MIC tube or well, the serial drug dilutions used in the test should be prepared 100 times the final test concentrations [4]. Despite this procedure, itraconazole and other agents will not remain solubilized at the highest drug concentrations upon further dilutions into the assay medium. The amount of turbidity at these concentrations may interfere with the interpretation of MIC endpoints. It is highly recommended to check the activity of the drug and the overall performance of the MIC method by a Quality Control (QC) practice to monitor the reliability of the test. QC isolates and expected MIC ranges are only available for the yeasts [4], but these QC isolates or stable mold isolates with known MICs can be used as controls.

Methods

Antifungal susceptibility testing for *A. fumigatus* has usually been performed by broth (macro- and microdilution) and agar dilution methods since 1956. To overcome the difficulties of evaluating the azoles and echinocandins, other less conventional approaches to susceptibility testing have been used, such as the measurement of a metabolic activity, e.g. the mycelial protein amount [5], glucose consumption [6], mitochondrial respiration [7], the inhibition of glucan [8] and ribonucleic acid syntheses, dry-weight increase [9] and the fungicidal (MFC) activity. Evaluations of fungal morphologic changes [8, 10] or determinations of a relative inhibition factor [11], the critical drug concentration by a disk diffusion assay [12], and of MICs by an agar diffusion [8] method also have been used as in vitro assays, since the early 1980s. More recently, the E test has been adapted in our laboratory for the testing of filamentous fungi, including *A. fumigatus*.

Test Media and pH

At least 17 solid or liquid media formulations have been chosen since 1956 as the test media for the determination of MICs for *A. fumigatus* [12–23]. Among these media, antibiotic medium 3 (M-3) and the chemically defined yeast nitrogen base (YNB) and RPMI-1640 (RPMI) media have been the formulations most commonly used [2, 7, 8–10, 13–35]. The effect of medium composition and pH on MIC values has been well documented. For example, unbuffered or acidic medium is unsuitable for antifungal susceptibility testing, owing to the adverse effect of a low pH on the activity of polyene and azole compounds. Amphotericin B MICs of 50 μg/ml have been reported for *A. fumigatus* when using YNB broth versus MICs of 1–5 μg/ml for the same strains in other media [9]. Reports of high amphotericin B MICs determined in unbuffered YNB have been controversial, e.g. amphotericin B MICs of 0.2–0.8 μg/ml [34] or above 4 μg/ml [9, 11, 26]. Van Cutsem et al. [23] also demonstrated the increase of itraconazole MICs in acidic medium, Sabouraud broth pH 5.6 MICs = 1–100 μg/ml and pH 7.4 MICs = 0.01–0.1 μg/ml for 87 isolates of *Aspergillus* spp.

RPMI, containing glutamine and MOPS buffer and without sodium bicarbonate at pH 7.0, is the NCCLS standard medium for the testing of yeasts [4]. This medium also gives reproducible MICs for most conidium-forming filamentous fungi with the established agents [28, 36] and voriconazole [27] as well as clinically relevant itraconazole MICs [2]. As for the yeasts, an increase of the glucose content of RPMI to 2% influences the growth of *A. fumigatus* and may facilitate an earlier reading of MICs [2, 25]. However, RPMI may not provide clinically relevant amphotericin B MICs for this and other fungal species. M-3 has been considered a better choice for determination of amphotericin B MICs, but lot variation has hampered the utility and evaluation of this medium [4]. Also, clinical failure to amphotericin B is more likely to be caused by host factors as very few resistant isolates with well documented clinical data are available for evaluation of this medium and other procedures for the detection of amphotericin B resistance of either yeasts or molds.

Inoculum Source, Preparation and Size

Sabouraud and potato dextrose agars have been the media most frequently used for the preparation of *A. fumigatus* inoculum suspensions. Although the parasitic form of *A. fumigatus* is the hyphal form, most MICs for this and other molds have been determined using non-germinated conidia or a mixture of hyphal and nongerminated conidial suspensions. MICs determined using non-germinated conidial suspensions reflect a measurement of the inhibition of conidial germination rather than the inhibition of hyphal growth. Since preparation of hyphal suspensions is cumbersome, a practical alternative is to induce conidial germination prior to MIC determination. Unfortunately, this alternative has been evaluated only in a few studies [7, 9, 14, 17] and both types of inocula have given comparable MIC endpoints [9, 14, 17]. But these studies have evaluated only the established drugs and a few strains of *A. fumigatus* without well-documented clinical data. Data are not available for the investigational agents.

The preparation of germinated conidia and hyphal growth usually requires 16–22 h of incubation at 30 °C in Czapek [14] and tryptic soy [7] broths. In our laboratory, we have obtained conidial germination after 8–9 h of incubation at 35 °C in either M-3 or RPMI media supplemented with 2% dextrose. Although complete conidial germination is obtained by this latter procedure, germinated conidia are not usually uniformly dispersed, a factor that can influence the MIC result. Furthermore, in order to identify which inoculum is clinically relevant, a standardized and reproducible procedure to prepare germinated conidial suspensions should be developed.

A wide range (10^2 to 10^7 CFU/ml) of inoculum sizes also have been utilized in the determination of MICs for *A. fumigatus*. The effect of inoculum size on MICs has been antifungal agent dependent. Significantly lower amphotericin B and fluorocytosine MICs were observed with an inoculum of 10^3 CFU/ml (0.05–0.1 and 5 μg/ml, respectively) than when the inoculum was 10^5 CFU/ml (0.5–1.0 and >100 μg/ml, respectively) [9]. In a more recent study, flucytosine MICs were 10-fold higher with 10^4 than with 10^2 CFU/ml inocula [29]; the effect was more moderate on itraconazole (five to sixfold increase) and amphotericin B MICs (two- to threefold increase). Collaborative studies by the NCCLS group have demonstrated that inoculum suspensions of approximately 10^4 CFU/ml prepared by following a standardized spectrophotometric procedure [37] produce reliable and reproducible MICs [28, 36].

Incubation Conditions

The incubation temperature and time also influence the MIC value. The temperatures used for fungal susceptibility testing have ranged from 22 to 37 °C. Most recent studies (after 1990) have used incubation temperatures between 35 and 37 °C, because these are the temperatures that best mimic what happens in the host. Kitahara et al. [9] evaluated the effect of 22 and 37 °C incubation temperatures on MIC endpoints of amphotericin B and flucytosine and found that the latter temperature produced more than fourfold higher MICs than the former temperature with both agents. With a few exceptions, most investigators have reported that 48 h amphotericin B MICs for *A. fumigatus* are equal to or below 2 μg/ml [10, 15, 32, 38]. A significant increase of amphotericin B values was noted by Gold et al. [19], in their first evaluation of the in vitro activity of this agent, when MICs were determined 48 (>40 μg/ml) rather than 24 h (1.6 μg/ml) after incubation. The inoculum for that study included a 'heavy' mixture of conidia and hyphal particles. It could be that what they observed was the combined effect of inoculum size and incubation time or only the former. A few other studies have reported amphotericin B MICs above 4 μg/ml when they were determined using unbuffered YNB medium as discussed above [9, 11, 26, 39] or after 72 h of incubation [14] for a few isolates of *A. fumigatus*; high MICs with this agent have been usually reported for other species of *Aspergillus* [31, 33].

Although no significantly different amphotericin B MICs have been reported for *A. fumigatus* between 24 h and 48 or 72 h of incubation, substantial differences have been demonstrated between 24 and 48 h itraconazole MICs when the inoculum sizes were $\leq 10^3$ CFU/ml [29]. Incubation time of 48 h for two microdilution methods [25, 28] and of 48–72 h for an agar dilution procedure [25] have provided clinically relevant itraconazole MICs (see below). However, the best incubation time for measuring the kinetics of glucose degradation was 72 h with amphotericin B and itraconazole [6].

Criterion for MIC Determination

Traditionally, the MIC endpoint was the minimum drug concentration that prevented visible growth. When testing fungicidal drugs such as amphotericin B, this criterion still applies. However, the azoles introduced the phenomenon of partial inhibition and by the traditional criterion high MICs of some of these agents did not correspond to their in vivo activity. The development of the NCCLS M-27 method for the yeasts [4] demonstrated that higher reproducibility of MICs was achieved when this partial inhibition was ignored for MIC determination of the azoles. High reproducibility of MICs was observed with the azoles against certain species of molds (including *A. fumigatus*) in two collaborative studies conducted by the NCCLS [28, 36] with the less-stringent criterion (prominent growth inhibition). Another approach is the evaluation of morphologic changes, which involves the determination of the lowest drug concentration that produces a change in morphology as seen by eye or the MEC [8].

Clinical Relevance

A reliable and reproducible method of MIC determination is essential to evaluate the clinical relevance of in vitro data. Although MICs can be useful in the selection and monitoring of the best therapeutic agent, the MIC is not the only, and may not be the best, predictor of the in vivo response to therapy. As discussed above, other numerous factors can influence this process, e.g. recovery of the neutrophil function may be essential for therapeutic response in neutropenic patients [40, 41]. For example, clinical response to itraconazole therapy [40, 41] and approximately 26% attributable itraconazole failure [40] were indiscriminately associated, using nonstandardized methods, with MICs of $\leq$1.0–6.3 μg/ml and MFCs of 0.8–12.5 μg/ml. Whilst low ($<$7 μg/ml) itraconazole serum concentrations were more often associated with lack of response than high MICs [40], especially in AIDS patients [42], serum levels did not always correlate with outcome [40, 43]. Lack of response to amphotericin B therapy for cerebral aspergillosis was correlated with an amphotericin B MIC of 4 μg/ml, while this patient had complete resolution of her lesions with 800 and 400 mg/day of itraconazole for nine months (itraconazole MIC of 1.0 μg/ml) [44]. Well-documented cases of the relationship between in vitro test results and patient response to therapy are rare, even for yeast infections; the availability of a reference method for the antifungal susceptibility testing of *A. fumigatus* and other molds would facilitate such documentation. Recently, certain testing guidelines have been proposed [28], which have improved interlaboratory consensus of MIC endpoints. In vitro resistance to itraconazole (MICs $>$16 μg/ml) has been identified by following these guidelines [unpubl. data] and similar testing conditions [2, 25] for two strains of *A. fumigatus* from 2 patients who did not respond to itraconazole therapy, despite adequate serum drug concentrations [2].

Although more recently animal studies have provided data regarding the relationship of MICs for *A. fumigatus* with treatment outcome with both established and investigational agents for infections in immunocompromised animals, these relationships can also be influenced by other factors. Itraconazole MICs of 0.2–2 μg/ml [2, 30, 32, 45], amphotericin B MICs and MFCs of

≤1.0–2 μg/ml, respectively [10, 32, 45–47], and a lyposomal nystatin MIC of 0.5 μg/ml [48] have been associated with response to treatment in several experimental aspergillosis infections. In contrast, lack of or poor response to treatment has been associated with itraconazole MICs of >16 μg/ml [2, 25, 45] and amphotericin B MICs and MFCs of 2 and >16 μg/ml, respectively [45]. The relationship with itraconazole has been dose and route of infection dependent. Animals infected with one of the two isolates of *A. fumigatus* with itraconazole MICs of >16 μg/ml had some response (20% mortality) at 75 mg/kg but not at 25 mg/kg itraconazole doses (80% mortality); the animals infected with the other strain had 80–100% mortality rates at both doses [2]. Resistance in the first strain was caused by reduced intracellular drug concentration and possibly it was mediated in the second strain by changes in the target enzyme [2], which may explain the different results. In another study, animals infected with isolates of *A. fumigatus* for which the itraconazole MIC was 2.2 μg/ml responded to itraconazole treatment when they were infected intravenously; however, the group infected intranasally did not respond to this treatment [49]. Itraconazole was not markedly effective when animals were infected as the latter group, despite lower itraconazole MICs and high levels in serum [13]. Additionally, a group of animals treated with itraconazole for experimental endocarditis were not cured, despite itraconazole MIC and MFC values of 0.38 and 0.78 μg/ml, respectively, for the infecting *A. fumigatus* isolate [22].

With the new triazoles, MICs of 0.01 to 0.39 μg/ml corresponded to effective therapy with voriconazole [22, 23], SCH 56592 [45] and ER-30346 [30] for experimental *A. fumigatus* infections in immunocompromised animals. A SCH 56592 MIC of 0.5 μg/ml correlated with an increase (100-fold) of CFU/organ counts, as compared with the results in a group of animals infected with an isolate that had an MIC of 0.01 μg/ml; both groups of animals received 25 mg/kg doses [25]. Despite its low in vitro activity (MICs of 8–16 μg/ml), the triazole D0870 was superior to amphotericin B and itraconazole in regard to animal mortality in a neutropenic model of pulmonary aspergillosis, when serum concentrations were above 4 mg/ml [46]. In a rat model of pulmonary aspergillosis [10], the activity of three pneumocandins correlated better with less traditional indicators of in vitro activity (MECs) and MICs by an agar diffusion procedure (MECs and MICs of <0.06–0.5 μg/ml) than with conventional MIC endpoints (>128 μg/ml).

Conclusions

The relationship of in vitro data with therapeutic response for *A. fumigatus* infections ought to be established based on the principle that antifungal resistance more likely would indicate therapeutic failure. Reproducible and reliable in vitro data are essential to evaluate correlations of in vivo response with MIC values and to establish interpretive breakpoints. With the development of a reference method for antifungal susceptibility testing isolates of *A. fumigatus*, MICs can play an important role when failure with a standard dose is observed. The reference method also is needed to obtain population distribution profiles of MIC endpoints, to identify potentially resistant isolates, and for the much needed correlations of MICs with clinical outcome.

Acknowledgements

I thank Julie Rhodes and David Ingroff for their help in the manuscript preparation.

References

1 Espinel-Ingroff A: History of medical mycology in the United States. Clin Microbiol Rev 1996;9: 235–272.

2 Denning DW, Venkateswarlu K, Oakley KL, Anderson MJ, Manning NJ, Stevens DA, Warnock DW, Kelly SL: Itraconazole resistance in *Aspergillus fumigatus*. Antimicrob Agents Chemother 1997;41:1364–1368.

3 Espinel-Ingroff A: Clinical relevance of resistance. Infect Dis Clin N Am 1997;11:929–944.

4 National Committee for Clinical Standards: Reference method for broth dilution antifungal susceptibility testing of yeasts. Tentative standard M27-A. Natl Com Clin Lab Standards, Villanova, 1997.

5 Yamada H, Kohno S, Maesaki S, Koga H, Kaku M, Hara K, Tanaka H: Rapid and highly reproducible method for antifungal susceptibility testing of *Aspergillus* species. J Clin Microbiol 1993;31:1009–1012.

6 Garrigues JC, de Fontenay GC, Linas MD, Lagente M, Seguela JP: New in vitro assay based on glucose consumption for determining itraconazole and amphotericin B activities against *Aspergillus fumigatus*. Antimicrob Agents Chemother 1994;38:2857–2862.

7 Jahn B, Martin E, Stueben A, Bhakdi S: Susceptibility testing of *Candida albicans* and *Aspergillus* species by a simple microtiter menadione-augmented 3-(4,5-dimethyl-2-thiazolyl)-2,5-diphenyl-^{2}H-tetrazolium bromide assay. J Clin Microbiol 1995;33:661–667,

8 Kurtz MB, Heath IB, Marrinan J, Dreikorn S, Onishi J, Douglas C: Morphological effects of lipopeptides against *Aspergillus fumigatus* correlate with activities against (1,3)-β-*D*-glucan synthase. Antimicrob Agents Chemother 1994;38:1480–1489.

9 Kitahara M, Seth VK, Medoff G, Kobayashi GS: Antimicrobial susceptibility testing of six clinical isolates of *Aspergillus*. Antimicrob Agents Chemother 1976;9:908–914.

10 Kurtz MB, Bernard EM, Edwards FF, Marrinan JA, Dropinski J, Douglas CM, Armstrong D: Aerosol and parenteral pneumocandins are effective in a rat model of pulmonary aspergillosis. Antimicrob Agents Chemother 1995;39:1784–1789.

11 Odds FC, Webster CE, Abbott AB: Antifungal relative inhibition factors: BAY 1-9139, bifonazole, butoconazole, isoconazole, itraconazole (R 51211), oxiconazole, Ro 14-4767/002, sulconazole, terconazole and vibunazole (BAY n-7133) compared in vitro with nine established antifungal agents. J Antimicrob Chemother 1984;14:105–114.

12 Bartizal K, Gill CJ, Abruzzo GK, Flattery AM, Kong L, Scott PM, Smith JG, Leighton CE, Bouffard A, Dropinski JF, Balkovec J: In vitro preclinical evaluation studies with the echinocandin antifungal MK-0991 (L-743,872). Antimicrob Agents Chemother 1997;41:2326–2332.

13 Allendoerfer R, Loebenberg D, Rinaldi MG, Graybill JR: Evaluation of SCH51048 in an experimental model of pulmonary aspergillosis. Antimicrob Agents Chemother 1995;39:1345–1348.
14 Bezjak V: Standardization of a hyphal inoculum of aspergilli for amphotericin B susceptibility testing. J Clin Microbiol 1985;21:509–512.
15 Brandsberg JW, French ME: In vitro susceptibility of isolates of *Aspergillus fumigatus* and *Sporothrix schenckii* to amphotericin B. Antimicrob Agents Chemother 1972;2:402–404.
16 Dixon D, Shadomy S, Shadomy HJ, Espinel-Ingroff A, Kerkering TM: Comparison of the in vitro antifungal activities of miconazole and a new imidazole, R41,400. J Infect Dis 1978;138:245–248.
17 Dupont B, Drouhet E: Early experience with itraconazole in vitro and in patients: Pharmacokinetic studies and clinical results. Rev Infect Dis 1987;9:S71–S76.
18 Espinel-Ingroff A, Shadomy S, Gebhart RJ: In vitro studies with R51,211 (itraconazole). Antimicrob Agents Chemother 1984;26:5–9.
19 Gold W, Stout HA, Pagano JF, Donovick R: Amphotericins A and B, antifungal antibiotics produced by a streptomycete. 1. In vitro studies. Antibiot Ann 1956;1955–1956:579–586.
20 Hoeprich PD, Merry JM: In vitro activities of two new antifungal azoles. Antimicrob Agents Chemother 1984;25:339–341.
21 Maesaki S, Kohno S, Kaku M, Koga H, Hara K: Effects of antifungal agent combinations administered simultaneously and sequentially against *Aspergillus fumigatus*. Antimicrob Agents Chemother 1994;38:2843–2845.
22 Martin MV, Yates J, Hitchcock CA: Comparison of voriconazole (UK-109,496) and itraconazole in prevention and treatment of *Aspergillus fumigatus* endocarditis in guinea pigs. Antimicrob Agents Chemother 1997;41:13–16.
23 Van Cutsem J, Van Gerven F, Janssen PAJ: Activity of orally, topically, and parenterally administered itraconazole in the treatment of superficial and deep mycoses: Animal models. Rev Infect Dis 1987;9:S15–S32.
24 del Poeta M, Schell WA, Perfect JR: In vitro antifungal activity of pneumocandin L-743,872 against a variety of clinically important molds. Antimicrob Agents Chemother 1997;41:1835–1836.
25 Denning DW, Radford SA, Oakley KL, Hall L, Johnson EM, Warnock DW: Correlation between in-vitro susceptibility testing to itraconazole and in-vivo outcome of *Aspergillus fumigatus* infection. J Antimicrob Chemother 1997;40:401–414.
26 Denning DW, Hanson LH, Perlman AM, Stevens DA: In vitro susceptibility and synergy studies of *Aspergillus* species to conventional and new agents. Diagn Microbiol Infect Dis 1992;15:21–34.
27 Espinel-Ingroff A: In vitro activity of the new triazole voriconazole (UK-109,496) against opportunistic filamentous and dimorphic fungi and common and emerging yeast pathogens. J Clin Microbiol 1998;36:198–202.
28 Espinel-Ingroff A, Bartlett M, Bowden R, Chin NX, Cooper C Jr, Fothergill A, McGinnis MR, Menezes P, Messer SA, Nelson PW, Odds FC, Pasarell L, Peter J, Pfaller MA, Rex JH, Rinaldi MG, Shankland GS, Walsh TJ, Weitzmann I: Multicenter evaluation of proposed standardized procedure for antifungal susceptibility testing of filamentous fungi. J Clin Microbiol 1997;35:139–143.
29 Gehrt A, Peter J, Pizzo PA, Walsh TJ: Effect of increasing inoculum sizes of pathogenic filamentous fungi on MICs of antifungal agents by broth microdilution method. J Clin Microbiol 1995;33:1302–1307.
30 Hata K, Kimura J, Miki H, Toyosawa T, Megumi M, Katsu K: Efficacy of ER-30346, a novel oral triazole antifungal agent, in experimental models of aspergillosis, candidiasis, and cryptococcosis. Antimicrob Agents Chemother 1996;40:2243–2247.
31 Kerkering TM, Espinel-Ingroff A: In vitro synergistic activity of itraconazole and 5-fluorocytosine against *Aspergillus fumigatus* and *Aspergillus flavus*; in Fromtling RA (ed): Recent Trends in the Discovery, Development and Evaluation of Antifungal Agents. Philadelphia, Prous, 1987, pp 61–70.
32 Murphy M, Bernard EM, Ishimaru T, Armstrong D: Activity of voriconazole (UK-109,496) against clinical isolates of *Aspergillus* species and its effectiveness in an experimental model of invasive pulmonary aspergillosis. Antimicrob Agents Chemother 1997;41:696–698.
33 Oakley KL, Moore CB, Denning DW: In vitro activity of SCH-56592 and comparison with activities of amphotericin B and itraconazole against *Aspergillus* spp. Antimicrob Agents Chemother 1997;41:1124–1126.

34 Schmitt HJ, Bernard EM, Andrade J, Edwards F, Schmitt B, Armstrong D: MIC and fungicidal activity of terbinafine against clinical isolates of *Aspergillus* spp. Antimicrob Agents Chemother 1988;32:780–781.
35 Yotsuji A, Shimizu K, Araki H, Fujimaki K, Nishida N, Hori R, Annen N, Yamamoto S, Hayakawa H, Imaizumi H, Watanabe Y, Narita H: T-8581, a new orally and parenterally active triazole antifungal agent: In vitro and in vivo evaluations. Antimicrob Agents Chemother 1997;41:30–34.
36 Espinel-Ingroff A, Dawson K, Pfaller M, Anaissie E, Breslin B, Dixon D, Fothergill A, Peter J, Rinaldi M, Walsh T: Comparative and collaborative evaluation of standardization antifungal susceptibility testing for filamentous fungi. Antimicrob Agents Chemother 1995;39:314–319.
37 Espinel-Ingroff A, Kerkering TM: Spectrophotometric method of inoculum preparation for the in vitro susceptibility testing of filamentous fungi. J Clin Microbiol 1991;29:393–394.
38 Shadomy S: In vitro antifungal activity of clotrimazole (Bay 5097). Infect Immunity 1971;4:143–148.
39 Ikemoto H: Treatment of pulmonary aspergilloma with amphotericin B. Arch Intern Med 1965; 115:598–601.
40 Denning DW, Lee JY, Hostetler JS, Pappas P, Kauffman CA, Dewsnup DH, Galgiani JN, Graybill JR, Sugar AM, Catanzaro A, Gallis H, Perfect JR, Dockery B, Dismukes WE, Stevens DA: NIAID Mycoses Study Multicenter trial of oral itraconazole therapy for invasive aspergillosis. Am J Med 1994;97:135–144.
41 Denning DW, Tucker RM, Hanson LH, Stevens DA: Treatment of invasive aspergillosis with itraconazole. Am J Med 1989;86:791–800.
42 Denning DW, Follansbee SE, Scolaro M, Norris S, Edelstein H, Stevens DA: Pulmonary aspergillosis in the acquired immunodeficiency syndrome. N Engl J Med 1991;324:654–662.
43 Lortholary O, Meyohas M-C, Dupont B, Cadranel J, Salmon-Ceron D, Peyramond D, Simonin D, Centre d-Informations et de Soins de l'Immunodéficience Humaine de l'Est Parisien: Invasive aspergillosis in patients with acquired immunodeficiency syndrome: Report of 33 cases. Am J Med 1993;95:177–186.
44 Sánchez C, Mauri E, Dalmau D, Quintana S, Aparicio A, Garau J: Treatment of cerebral aspergillosis with itraconazole: Do high doses improve the prognosis? Clin Infect Dis 1995;21:1485–1487.
45 Oakley KL, Morrissey G, Denning DW: Efficacy of SCH-56592 in a temporarily neutropenic murine model of invasive aspergillosis with an itraconazole-susceptible and an itraconazole-resistant isolate of *Aspergillus fumigatus*. Antimicrob Agents Chemother 1997;41:1504–1507.
46 Denning DW, Hall L, Jackson M, Hollis S: Efficacy of D0870 compared with those of itraconazole and amphotericin B in two murine models of invasive aspergillosis. Antimicrob Agents Chemother 1995;39:1809–1814.
47 Patterson TF, Miniter P, Andriole VT: Efficacy of fluconazole in experimental invasive aspergillosis. Rev Infect Dis 1990;12:S281–S285.
48 Wallace TL, Paetznick V, Cossum PA, Lopez-Berestein G, Rex JH, Anaissie E: Activity of liposomal nystatin against disseminated *Aspergillus fumigatus* infection in neutropenic mice. Antimicrob Agents Chemother 1997;41:2238–2243.
49 Graybill JR, Ahrens J: Itraconazole treatment of murine aspergillosis. J Med Vet Mycol 1985;23: 219–223.

Dr. A. Espinel-Ingroff, Division of Infectious Diseases,
Medical College of Virginia/Virginia Commonwealth University,
Medical Mycology Research Laboratory, Box 980049, 1101 E. Marshall Street,
Richmond, VA 23298-0049 (USA)
Tel. +1 (804) 828 9711, Fax +1 (804) 828 3097

Brakhage AA, Jahn B, Schmidt A (eds): *Aspergillus fumigatus.*
Contrib Microbiol. Basel, Karger, 1999, vol 2, pp 149–166

Transformation Systems of *Aspergillus fumigatus*

New Tools to Investigate Fungal Virulence

Christophe d'Enfert [a], *Gerhard Weidner* [b], *Pieternella C. Mol* [a], *Axel A. Brakhage* [b]

[a] Laboratoire des Aspergillus, Institut Pasteur, Paris, France;
[b] Institut für Mikrobiologie und Genetik, Technische Universität Darmstadt, Deutschland

Over the past two decades molecular genetics has become one of the most productive areas in the investigation of the mechanisms of microbial pathogenesis and in the development of novel strategies leading to the identification of antimicrobial drugs. Molecular genetics offers not only approaches for the detailed characterization of gene products that appear to be key components of the infection process but also selection strategies that combine classical genetics and molecular biology to identify virulence determinants of microbial pathogens [1]. The study of a microbial pathogen with molecular genetic techniques requires the development of a gene transfer system that allows for the manipulation of its genome and the identification of genetically defined mutants of this particular organism. Gene transfer systems consist of a vector carrying a selectable marker and of a transformation procedure to introduce this vector into the host. Transformation of *Aspergillus fumigatus* has long been hampered by the poor development of classical genetics in this species and consequently of auxotrophs that could serve as recipient strains for vectors carrying the corresponding selectable marker. The development in related *Aspergillus* species of transformation systems taking advantage of dominant selectable markers based on bacterial antibiotic resistance genes paved the way for the implementation of similar systems in *A. fumigatus* [2, 3]. DNA-mediated transformation of *A. fumigatus*, pioneered by the groups of Holden and coworkers [4, 5] and of Latgé and Monod and coworkers [6, 7], is now routinely used in several laboratories to investigate the role of specific gene products in *A. fumigatus* virulence and/or biology. Here we will review (1) the selectable markers that can be used in *A. fumigatus*; (2) the transformation procedures that are available to introduce genetic material into *A. fumigatus*; (3) the modifications of the genome that can be obtained by DNA-mediated transformation, and (4) the strategies which are now facilitated by the availability of transformation systems and should reveal those gene products that are essential for *A. fumigatus* growth and pathogenicity.

Selectable Markers

A prerequisite for the establishment of a transformation system is the availability of a selectable marker that will enable the identification of strains that have received exogenous recombinant DNA and are therefore likely to retain an appropriate modification of their genome. Development of transformation systems in fungi has mostly relied on the use of strains with recessive auxotrophic mutations that could be complemented by the introduction of the wild-type copy of the *auxotrophic marker*. An alternative strategy has been provided by the sensitivity of some filamentous fungi to various drugs and the consequent use of *dominant selectable markers* that confer resistance to a specific drug when introduced into the fungal host [3].

Dominant Selectable Markers

A wide variety of filamentous fungi including *Aspergillus* species are sensitive to several antibiotics, e.g. hygromycin and phleomycin. The availability of bacterial resistance genes whose products counteract the toxic effects of these antibiotics in bacteria allowed the design of cassettes comprising a resistance gene whose natural expression signals have been replaced by fungal promoter and terminator regions, in particular those of the *A. nidulans gpdA* or *trpC* genes [2]. These cassettes have been successfully used to assess the entry of genetic material into various fungal species [3]. Sensitivity of *A. fumigatus* to both hygromycin and phleomycin allowed the use of these cassettes as selectable markers in transformation experiments. As can be seen from table 1, hygromycin resistance has been widely used as selectable marker in the construction of knock-out mutant strains of *A. fumigatus* while phleomycin resistance has been used only in several instances, in particular when mutants carrying two gene knock-outs were sought [4–17]. The favored use of hygromycin resistance may in part reflect variations in sensitivity of *A. fumigatus* strains to phleomycin and the consequent difficulty that is observed in the selection stage when random natural isolates of *A. fumigatus* are transformed [Beauvais, Mouyna and Paris, pers. commun.].

Selection of transformation events in filamentous fungi by means of drug resistance has also relied on the use of additional dominant selectable markers

Table 1. Strategies used to construct defined knock-out mutants in *A. fumigatus*

Gene	Function	Selectable marker[1]	Transformation procedure	Integration event	Reference
alp	extracellular alkaline protease	Hyg^R	protoplast	gene disruption	4
		Hyg^R	protoplast	gene replacement	6
		Phl^R	protoplast	gene replacement	8
mep	extracellular metalloprotease	Phl^R	protoplast	gene replacement	9
mepB	intracellular Zn-oligopeptidase	Hyg^R	protoplast	gene disruption	10
pep	aspergillopepsin	Phl^R	protoplast	gene replacement	11
res/aspf1	restrictocin	Hyg^R	protoplast	gene disruption	5
		Hyg^R	protoplast	gene replacement	7
chsC	chitin synthase	Hyg^R	protoplast	gene replacement	12
chsD	chitin synthase	Hyg^R	protoplast	gene replacement	13
chsE	chitin synthase	Hyg^R	protoplast	gene replacement	14
chsG	chitin synthase	Phl^R	protoplast	gene replacement	12
arp1	scytalone dehydratase	Hyg^R	protoplast	gene replacement	15
rodA	hydrophobin, component of the	Hyg^R	protoplast	gene replacement	16
	conidial rodlet layer	*A. niger pyrG*	protoplast, electroporation	gene replacement	26 C.E., unpubl. data
cat1	catalase	Hyg^R	protoplast	gene replacement	17
pabaA	p-aminobenzoic acid biosynthesis	Hyg^R	protoplast, electroporation	gene replacement	Brown, pers. commun.

[1] Hyg^R: Hygromycin resistance; Phl^R: phleomycin resistance.

other than hygromycin or phleomycin resistance [3]. Of particular interest are genes that can confer resistance to benomyl [18–20] or phosphinothricin [21, 22] and that have been used successfully in several fungal species. Although this has not been evaluated yet, they are likely to provide additional convenient markers for the transformation of *A. fumigatus.*

Auxotrophic Markers

Dominant selectable markers have a major advantage as they can be readily used to select transformants of natural isolates of *A. fumigatus.* However, they present several drawbacks. First, as outlined above, only a limited number of dominant selectable markers is available thus precluding the construction of multiple-disrupted strains that may be required when gene families or complex cellular processes are studied. Second, they lack the versatility provided by several auxotrophic markers which can be used to precisely modify genes of *A. fumigatus* using strategies that have been developed in other fungal species such as *Saccharomyces cerevisiae* and *A. nidulans* [23–25]. In this regard, auxotrophic markers that can be both selected by complementation of a recessive mutation and counter-selected by the sensitivity of prototrophs to a specific drug are of particular interest for the manipulation of fungal genomes (see 'Genetic manipulations').

Because the genus *Aspergillus* belongs to the deuteromycetes which lack a sexual cycle, classical genetics of *A. fumigatus* has been poorly developed and very few auxotrophic mutants have been characterized. Strategies are nevertheless available to positively select spontaneous auxotrophs that have defined recessive mutations and these have been successfully applied to *A. fumigatus*. In particular, *pyrG* mutants that lack orotidine-5′-phosphate (OMP) decarboxylase activity and hence are auxotrophs for uridine or uracil have been identified among strains that are resistant to 5-fluoro-orotic acid (5-FOA) [26, 27]. Similarly, *niaD* mutants that lack nitrate reductase activity and hence cannot grow on media containing nitrate as the nitrogen source have been identified among strains that are resistant to chlorate [6, 28].

The availability of *A. fumigatus pyrG* strains and of OMP decarboxylase-encoding genes from related *Aspergillus* species provided the basis for the development of a transformation system that uses the *A. niger pyrG* gene [29] as selectable marker and a mutant *A. fumigatus pyrG1* isolate as recipient [27]. More recently, the *A. fumigatus pyrG* gene has been cloned and successfully used to select transformants of an *A. fumigatus pyrG1* strain thus allowing novel genetic manipulation strategies in *A. fumigatus* [30] (see 'Genetic manipulations'). One drawback in the implementation of this transformation system in natural isolates of *A. fumigatus* lies in the necessary identification of a *pyrG* mutant among 5-FOA-resistant clones. This can be achieved with either an enzymatic assay or a complementation test that involves the transformation of all putative mutants with a cloned *pyrG* gene. However, we recently characterized the *pyrG1* mutation. Transformation of wild-type isolates with the cloned *pyrG1* allele should yield 5-FOA-resistant transformants that carry the *pyrG1* mutation and are suitable for further genetic manipulations [30, 31].

As mentioned above, *A. fumigatus niaD* mutants are available and could be transformed using *niaD* genes from related *Aspergillus* species thus providing an additional transformation system for *A. fumigatus* with properties similar to that of the *pyrG*-based transformation system [32]. The cloning of several *A. fumigatus* auxotrophic markers (*trpC, HIS3, LEU2, TRP1* and *pabaA*) by functional complementation of corresponding mutations in *A. nidulans* or *S. cerevisiae* [33, 34; Brown and Aufauvre-Brown, pers. commun.] should provide the basis for the construction of defined auxotrophs of *A. fumigatus* and subsequent development of novel transformation systems for this species. However, although *A. fumigatus pabaA* mutants have been obtained, their use as recipient strains in transformation experiments has not been evaluated to date [Brown and Aufauvre-Brown, pers. commun.].

A limitation to the use of auxotrophic markers is the reduced virulence of some *A. fumigatus* auxotrophs, e.g. *pyrG* mutants [35] and *pabaA* mutants [36, 37]. However, since transformants are selected that have regained prototrophy, they are likely to simultaneously recover virulence as is the case for *A. fumigatus pyrG* mutants [35]. *A. fumigatus* strains with multiple auxotrophies may nevertheless not be suitable for further virulence testing if transformed by a single auxotrophic marker. A general concern in the application of transformation systems using auxotrophic markers results from the need to identify and/or construct auxotrophs prior to implementing a transformation system in a particular natural isolate of *A. fumigatus*. In this respect, the use of a single wild-type isolate and of related auxotrophs by the different laboratories involved in the study of *A. fumigatus* would be beneficial for both strain availability and data consistency.

Transformation Procedures

Establishment of transformation systems in *A. fumigatus* has relied on methods developed in the 1980s for other filamentous fungi, in particular *A. nidulans* and *A. niger*. As a consequence, protoplasts of *A. fumigatus* prepared with commercially available lytic enzymes (Novozym 234, Glucanex) have been favored as recipients in transformation experiments that use PEG/ $CaCl_2$ to promote the entry of DNA into cells [38]. Regeneration of protoplasts is conducted in osmotically stabilized selective media that contain the drug corresponding to the dominant selectable marker carried by the transforming DNA [4, 6, 8, 9; see 39 and 40 for detailed protocols] or that lack the nutrient corresponding to the nutritional requirement of the recipient strain [27]. Transformation efficiencies vary between 2 and 100 transformants/μg of transforming DNA depending on the nature of the ensuing integration event (see 'Genetic manipulations').

Although protoplast transformation has been used successfully in many instances to construct mutants of *A. fumigatus* [4–17], it is technically strenuous and somewhat unreliable because the production and viability of protoplasts and their competency for DNA uptake or integration are not highly reproducible. Using a protocol developed for *A. nidulans* [41], we have recently established a transformation procedure for introducing DNA into germinating conidiospores of *A. fumigatus* by electroporation [30]. Major advantages of this procedure are the absence of enzymatic treatment of the germinating spores prior to electroporation and the direct selection of transformants on standard selective media without the need for osmotic stabilizers. Furthermore, transformation efficiencies are 2- to 10-fold higher than those obtained with the protoplast-PEG/$CaCl_2$ procedure [30; Brown and Aufauvre-Brown, pers. commun.]. While these characteristics would make electroporation a method of choice for the transformation of *A. fumigatus*, we and others have noticed that ectopic integration of the transforming DNA is favored [C.E., unpubl. data; Brown and Aufauvre-Brown, pers. commun.]. This precludes the use of electroporation when site-directed integration of the DNA is sought unless strategies are implemented that selectively target transforming DNA at a particular location in the genome or allow the rapid identification of appropriate transformants (see 'Genetic manipulations').

Genetic Manipulations

Autonomously Replicating Plasmids

In most instances DNA-mediated transformation of *A. fumigatus*, as in other *Aspergillus* species, results in the integration of the transforming DNA into the genome of the recipient cell, hence providing a tool to construct genetically defined stable mutants of this species. Transformation of *A. fumigatus* using autonomously replicating plasmids carrying the AMA1 plasmid replicator [42] has also been achieved. In this case, transformation efficiencies are dramatically increased and the plasmid behaves autonomously and is unstable under nonselective conditions [d'Enfert, Gehringer, Langfelder and Brakhage, unpubl. obs.]. These plasmids provide a convenient monitoring tool during the establishment of transformation procedures because of their ability to transform at high frequency. More importantly, strategies developed in *A. nidulans* and other filamentous fungi that take advantage of these plasmids in the cloning of genes by functional complementation or the study of gene regulation should be applicable in *A. fumigatus* [42]. However, the use of these plasmids remains somewhat limited until strategies are developed to control their copy number and prevent their loss under nonselective conditions.

Cloning of Genes or Characterization of Mutations by Functional Complementation

Very few mutants of *A. fumigatus* obtained by physical or chemical mutagenesis have been characterized in detail. Transformation with genes from other *Aspergillus* species has been used to define the genetic basis of specific phenotypes. In particular, complementation of a white-spored strain of *A. fumigatus* has been obtained by transformation with the *wA* gene of *A. nidulans* demonstrating that the white-spored phenotype was likely due to a mutation in the *A. fumigatus wA* homologue [33]. Similarly, identification of *A. fumigatus pyrG* mutants among 5-FOA-resistant strains has relied on a transformation test with the *A. niger pyrG* gene [27]. Recently, DNA-mediated transformation with a genomic library of *A. fumigatus* constructed in a cosmid that carries a hygromycin resistance marker has been used successfully to complement a genetically uncharacterized mutation and identify a gene that is involved in conidial pigmentation in *A. fumigatus* [15]. This shows that, as in many other species, DNA-mediated transformation provides an efficient tool to identify *A. fumigatus* genes by functional complementation when a mutant phenotype is scored easily. As mentioned above, this strategy is thought to be of limited value in *A. fumigatus* since the lack of a sexual cycle precludes a careful genetic study of the mutations that underly a mutant phenotype. However, a parasexual cycle has been established for *A. fumigatus* and ultimately allows the genetic characterization of mutants [43].

DNA-Mediated Transformation as a Tool for Insertional Mutagenesis

Genetic manipulation of *A. fumigatus* following DNA-mediated transformation is possible since transforming DNA becomes integrated by homologous recombination (see below). However, the frequency of homologous recombination in *A. fumigatus* as in other filamentous fungi is low and many transformants result from ectopic integration (fig. 1A). If the transforming DNA molecule does not carry *A. fumigatus* sequences, integration occurs randomly in the genome thus providing a basis for the establishment of a collection of insertional mutants of *A. fumigatus*. If a particular phenotype can be associated with a single insertion in the genome, strategies are then available to identify the genomic regions that flank this insertion and hence the gene whose disruption results in the given phenotype.

It has been noticed that in several fungi the combined use of linearized DNA and addition of a particular restriction enzyme in the transformation mixture increases transformation efficiency and promotes single-copy integration of the transforming DNA at genomic sites that correspond to the restriction enzyme recognition sequence [44, 45]. This process, referred to as restriction enzyme-mediated integration (REMI; fig. 1B), has been used to construct collections of insertional mutants in several filamentous fungi and has allowed the identification of genes involved in toxin biosynthesis or pathogenicity [46, 47]. REMI has also been employed to construct single-copy insertional mutants of *A. fumigatus* using protoplast transformation and a linearized vector carrying a hygromycin resistance gene. The mutants obtained in this manner await further characterization in pathogenicity tests [37]. Data reported for *A. fumigatus* as well as other fungi suggest that only a subset of restriction enzymes added in defined amounts is able to promote REMI in a given organism, thus limiting the number of random mutations that can be created by this process. In this regard, transformation by electroporation provides a more valuable tool – provided that illegitimate recombination occurs randomly – to generate a large collection of independent insertional mutants of *A. fumigatus* because it gives higher transformation efficiencies and as a rule results in single-copy integrations independently of the addition of restriction enzymes [30; Brown and Aufauvre-Brown, pers. commun.].

Targeted Mutagenesis of Specific Genes

Recombination between genomic sequences and homologous sequences carried by recombinant DNA molecules has been now used extensively to construct *A. fumigatus* strains that have a defined mutation at a particular locus. While homologous recombination can occur when fragments as small as 300 bp are used [13], it is with low efficiency and larger regions are required (ca. 1 kb). Although illegitimate recombination represents a major constraint in the analysis of transformants obtained when using constructs aimed to integrate by homologous recombination, this latter process occurs at a frequency sufficient to allow for the identification of appropriate mutants among a relatively low number of transformants (20–50) with various techniques as enzymatic assay, Western blotting, PCR analysis on intact spores or purified genomic DNA and Southern analysis of genomic DNA [39, 40, 48]. Several strategies are available to create gene knock-outs or site-specific mutations in *A. fumigatus*:

Gene Disruption versus Gene Replacement in the Generation of Gene Knock-Outs. Most of the mutations that have been created to date through DNA-mediated transformation are knock-outs in genes that encode proteins postulated to participate in *A. fumigatus* virulence (proteases, toxins, detoxifying enzymes, conidial morphology; table 1) or in the morphogenesis of the fungus (chitin synthases; table 1). Two strategies have been used that are illustrated in Fig. 1C (*gene disruption*) and Fig. 1E (*gene replacement*). In *gene disruption* experiments, an internal fragment of the gene of interest is cloned in a vector that carries a selectable marker. Upon transformation, homologous integration of the resulting plasmid via a single crossover results in two mutant copies of the gene separated by the selectable marker, both thought to be nonfunctional since the first encodes a truncated peptide and the second lacks the signals required for proper expression of the gene as well as the 5′-coding region of

Fig. 1. Strategies used to manipulate the genome of *A. fumigatus* by DNA-mediated transformation. The *A. fumigatus* genome is shown as a thick line, with a gene of interest shown as a black box, while vector DNA is shown as a thin line. Site-specific mutations are indicated by a white circle. Selectable markers are indicated as a white box or a grey box if they can be counter-selected. Repeats in blasters are shown by a boxed arrow. Illegitimate recombination has been indicated by dashed lines while homologous recombination is shown as plain lines. X = Cleavage site for restriction enzyme X.

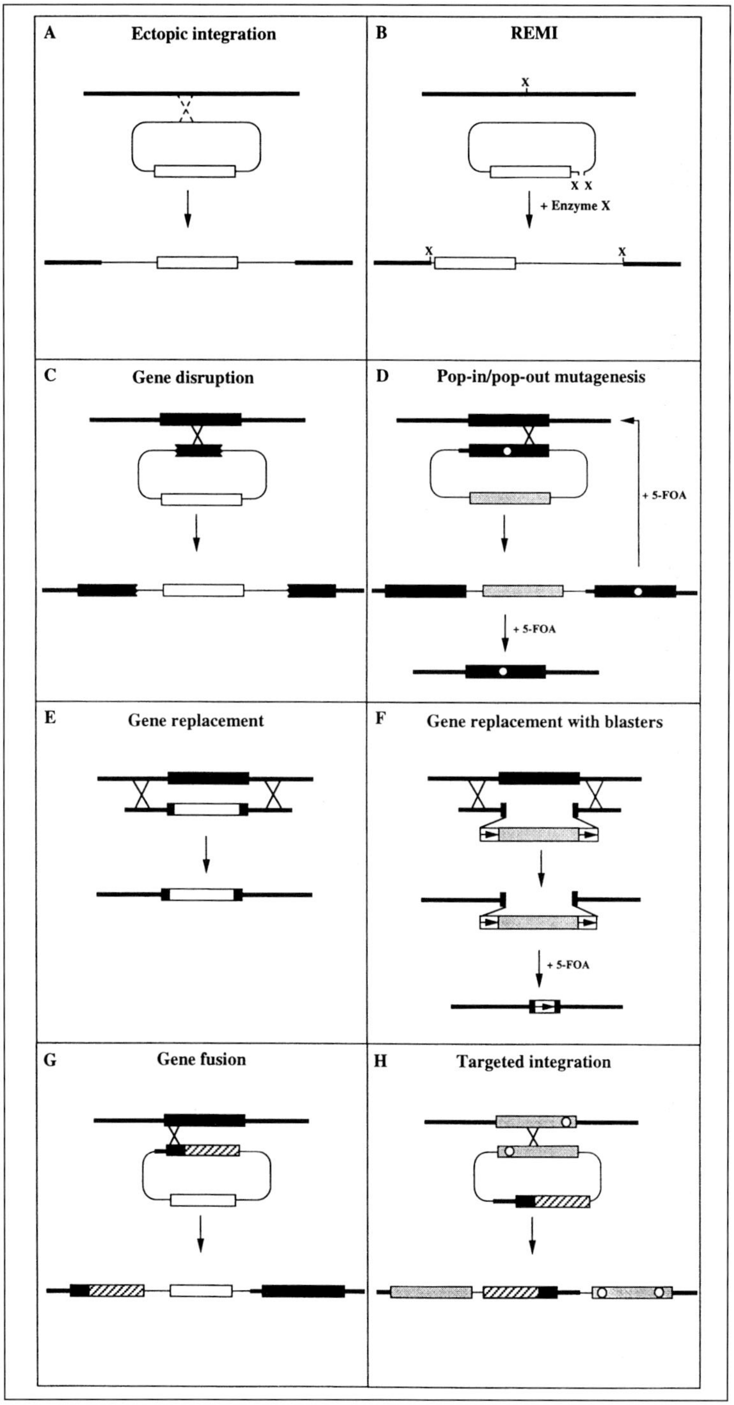
A
Ectopic integration
B
REMI
X
X X
+ Enzyme X
X
X
C
Gene disruption
D
Pop-in/pop-out mutagenesis
+ 5-FOA
+ 5-FOA
E
Gene replacement
F
Gene replacement with blasters
+ 5-FOA
G
Gene fusion
H
Targeted integration

the gene. In *gene replacement* experiments, a selectable marker is inserted within a gene, with or without the simultaneous deletion of a part of its coding region. Upon transformation, homologous recombination by a double crossover event between the genomic sequences and the regions flanking the selectable marker results in replacement of the wild-type allele by the mutant gene.

As can be seen in table 1, gene replacement has been used more extensively than gene disruption and this may be for the following reasons: (1) gene disruption yields a truncated peptide that may still be functional, a problem that is alleviated in gene replacement experiments by cloning the selectable marker in the 5′-coding region of the gene or by deleting part of the coding sequence; (2) gene disruption involves the use of an internal fragment of the gene and the low efficiency of homologous recombination observed with short DNA regions may preclude the use of gene disruption when genes of relatively small size are mutated, and (3) although this has not been reported yet, gene disruption creates a tandem repeat that is potentially unstable under nonselective conditions

Construction of Strains with Multiple Gene Knock-Outs. Construction of *A. fumigatus* strains with multiple gene knock-outs involves the iteration of the strategies described above using a different selectable marker in each round of transformation. This strategy has been employed to construct *A. fumigatus* strains that simultaneously lack the secreted ALP alkaline protease and MEP metallo-protease, restrictocin and the ALP protease, or the CHSC and CHSG chitin synthases, one gene being interrupted by the hygromycin resistance gene and the second by the phleomycin resistance gene [8, 9, 12]. As mentioned above, few selectable markers have been evaluated for *A. fumigatus* transformation, thus limiting the number of gene knock-outs that can be successively introduced in a single strain. To circumvent this, a strategy has been developed that uses a unique selectable marker in successive transformation experiments. This strategy relies on a tripartite selectable marker, referred to as a *pyrG*-blaster, that contains the *A. niger pyrG* gene flanked by a direct repeat that encodes the neomycin phosphotransferase gene (*neo*) of transposon Tn5 [27; fig. 1F]. Following the identification of a prototrophic transformant of a *pyrG* strain of *A. fumigatus* with the appropriate gene replacement, recombination between the two elements of the direct repeat is induced in the presence of 5-FOA, resulting in the excision of the *A. niger pyrG* gene and the production of a strain that has regained uridine/uracil auxotrophy and maintains a gene knock-out because of the persistence of only one element at the site of insertion of the *pyrG*-blaster (fig. 1F). This strain can be used in a second round of transformation that uses *pyrG* as selectable marker. 5-FOA dependent recombination between the two *neo* genes has been shown to occur at frequencies as high as 10^{-4} in *A. fumigatus* as well as in *A. nidulans* and *A. oryzae* [27, 49; Abe, pers. commun.]. Apart from allowing the construction of strains with multiple gene knock-outs that may be useful when complex, cellular processes or gene families (e.g. chitin synthases [50] or extracellular proteases [9, 11, 51–54; see Monod et al., this vol.]) are studied, this strategy is of particular interest because it has the potential to generate strains with one or several knock-outs that have identical genetic and phenotypic backgrounds and hence are more suitable for further comparisons.

Pop-In/Pop-Out Mutagenesis. Strategies described in the two previous paragraphs result in *A. fumigatus* strains that are mutated through gene knock-outs. Although this approach has proved useful in the study of proteins that have presumed to participate in *A. fumigatus* morphogenesis or virulence, it only allows to compare on and off situations and does not bring any insight in the structure-function relationships for a specific protein. Furthermore, knock-outs in genes that are involved in essential cellular processes are often lethal. In this case, protoplast transformation is thought to result in the formation of a viable heterokaryon with wild-type and mutant nuclei that is only resolved into viable wild-type and nonviable mutant homokaryons following conidiogenesis [55]. This should permit a characterization of the early phenotypes associated with a knock-out mutation in an essential gene. Pop-in/pop-out mutagenesis provides an alternative for the replacement of a wild-type gene by mutant alleles (deletion, insertion, site-specific mutations) in a more controlled manner compared with standard gene knock-out strategies [23]. This method is now applicable in *A. fumigatus* [27]. As shown in figure 1D, a specific alteration is introduced into the gene of interest cloned in a *pyrG*-based integrating vector. Upon transformation of an *A. fumigatus pyrG* strain, prototrophic transformants are obtained that contain the wild-type gene and the mutant copy flanking the plasmid sequences (pop-in). In the presence of 5-FOA, recombination between the wild-type and mutant copies will produce strains that have either the wild-type copy of the gene, if recombination occurs on the same side of the mutation as during integration of the transforming vector, or the mutant copy, if recombination occurs on the opposite side (pop-out). This strategy is of particular interest since it allows construction of site-specific mutants with a particular phenotype, e.g. thermosensitivity, as well as the evaluation of the consequences of a gene knock-out since lethality will only result in the production of wild-type strains following selection on 5-FOA.

Gene Fusions

Understanding the mechanisms by which gene expression is controlled in *A. fumigatus*, in particular in the context of the infection process, will undoubtedly contribute to improve our knowledge of this fungus. In this regard, gene fusions between promoter sequences of an *A. fumigatus* gene and a reporter gene represent a major genetic tool to study the expression of genes in a particular environment, and to identify sequences that contribute to gene expression and genes that participate in the control of gene expression.

Two strategies are available to introduce gene fusions in the genome of *A. fumigatus*. First, an integrative vector that carries an in-frame fusion between the 5′ region of a gene of interest and a reporter gene is used. This construct will integrate by homologous recombination therefore creating a genomic gene fusion expressed under the control of the full-length promoter region of the gene (fig. 1G). It is followed by the vector sequences and a functional or nonfunctional copy of the gene according to the length of the 5′ non-coding region which is present in the transforming DNA (fig. 1G). This approach has been used successfully by Smith et al. [8] to create a fusion between the *A. fumigatus ALP* gene and the *Escherichia coli lacZ* gene encoding β-galactosidase and to demonstrate expression of the *ALP* gene during colonization of the lung by *A. fumigatus* in a mouse model of invasive aspergillosis. The *E. coli uidA* gene encoding β-glucuronidase can also serve as a reporter in *A. fumigatus* [Mol, unpubl. data]. It is likely that, as shown in *A. nidulans* [56], the *Aequorea victoria* green fluorescent protein (GFP) gene can be used as well thus providing a tool to study both gene expression and protein localization in *A. fumigatus*.

However, the above strategy remains of limited use since it only results in gene fusions that are expressed under the control of the full-length promoter region of the gene of interest thus precluding the characterization of sequences that are controlling gene expression. This limitation can be alleviated by targeting the integration of gene fusions constructed in vitro at a preselected chromosomal location, in particular the locus of the marker used to select transformants. Tools to implement this strategy have been recently developed and rely on the use of a mutant allele (*pyrG2*) of the *A. fumigatus pyrG* gene: upon transformation of an *A. fumigatus pyrG1* strain with an integrative vector carrying both the *pyrG2* allele and a gene fusion, prototrophic transformants result from the restoration of a functional *pyrG* gene. This can occur by two mechanisms: (i) gene conversion of the *pyrG1* mutation, or (ii) integration of the transforming DNA via a single crossover between the *pyrG2* and *pyrG1* mutation [30] (fig. 1H). Using this approach, we have shown that up to 45% of the transformants that had integrated a plasmid carrying the *pyrG2* allele resulted from its integration at the chromosomal *pyrG* locus [30]. This latter strategy, in addition to allowing the introduction of gene fusions into the genome of *A. fumigatus*, enables the introduction of mutant genes conferring a dominant phenotype or of numerous copies of a single gene, a process that is also useful to study gene function and regulation [3, 25, 57].

DNA-Mediated Transformation as a Tool to Investigate the Molecular Mechanisms of *Aspergillus fumigatus* Virulence

Most of the techniques described in the previous section have been applied to understand whether specific *A. fumigatus* gene products that had been identified using biochemical or molecular biology approaches and postulated to participate in cellular morphogenesis or virulence of the fungus are indeed required during the infection process. When knock-out mutants have been evaluated in an animal model of invasive aspergillosis, little or no differences to a wild-type strain have been observed suggesting that these genes, although they may somehow contribute to *A. fumigatus* ability to colonize the host tissues, are not essential to virulence. It is clear that numerous genes remain to be identified that contribute to *A. fumigatus* virulence. Studies conducted with bacterial pathogens show that several techniques relying on the introduction of DNA into these organisms can be utilized to identify on a random basis genes that are predominantly expressed during infection and required for virulence [1, 58, 59]. Three strategies have been developed that imply the construction of a collection of genetically modified pathogens:

(1) *Signature tagged mutagenesis* (STM) is a negative selection scheme that involves the construction of a collection of insertional mutants (e.g. using transposon mutagenesis) that can be differentiated by an individual tag and that are tested in pools for virulence in an animal model [60]. This technique allows the identification of avirulent mutants among a large number of virulent strains within a single animal. It opens up the possibility to identify factors that are essential for the progression of the disease in the host environment.

(2) *In vivo expression technology* (IVET) was developed to identify promoter regions that are efficiently expressed during infection [61]. It uses gene fusions between random genomic regions and a promoterless gene whose expression is essential for virulence (e.g. auxotrophic marker genes). Pools of strains, each carrying an integrated independent gene fusion, are assayed in an animal model of infection, allowing the identification of strains that have regained virulence because the cloned promoter is functional during colonization. Promoter regions identified in this manner and that are not expressed in vitro correspond to genes that are specifically expressed in the host and are likely to participate in the infection process.

(3) *Differential fluorescence induction* (DFI) combines the use of gene fusions between random genomic regions and the GFP gene and of fluorescence-activated cell sorter (FACS) assisted analysis to identify promoter regions that show a differential pattern of expression in vivo (animal model or cell culture model) and in vitro (standard laboratory media) [62].

The application of STM, IVET and DFI to the study of *A. fumigatus* virulence is likely to bring a wealth of information on the factors that contribute to the fitness of the fungus in the host environment. Several adaptations will be needed since some of the techniques used to modify bacterial genomes are not applicable in fungi. Furthermore, transformation efficiencies are highly influenced by the nature of the recombination event that is promoted and this may limit the use of some strategies when generating large collections of genetically modified *A. fumigatus*.

As mentioned above, STM has been developed in bacteria by taking advantage of transposon mutagenesis for the generation of a collection of insertional mutants [60]. Transposon mutagenesis in filamentous fungi has not been developed yet and other techniques are needed to generate a collection of insertional mutants of *A. fumigatus*. In this regard, Brown et al. have combined the process of ectopic integration (see above) with STM: 96 plasmids, each carrying a unique tag and the hygromycin resistance gene, have been used individually in a REMI transformation procedure and in an electroporation procedure to generate ca. 3,500 transformants that were grouped in pools of 96 mutants, each carrying a single insertion with a different tag [37; Brown, pers. commun.]. These pools can be subsequently used in a mouse model of invasive aspergillosis to distinguish those strains that are unable to colonize lung tissues. Reconstitution experiments using a known avirulent mutant of *A. fumigatus* within a pool of random mutants have been used to validate this technique [37] and the identification of avirulent *A. fumigatus* strains within the collection of STM mutants is still in progress [Brown and Aufauvre-Brown, pers. commun.]. However, the STM approach may, in particular, suffer from the lack of consistency of the isolates recovered from a single animal inoculated with numerous *A. fumigatus* strains [Latgé, pers. commun.].

IVET and DFI involve the construction of genomic libraries in integrative vectors that contain a reporter gene – an auxotrophic marker gene essential for survival in host tissue or an antibiotic resistance gene for IVET or the GFP gene for DFI – and their subsequent integration at a specific locus in the genome. Since *pyrG* and *pabaA* mutants of *A. fumigatus* have reduced virulence [35, 37], the *pyrG* or *pabaA* genes are suitable for IVET in *A. fumigatus*. Similarly, GFP has been used successfully in *A. nidulans* [56] and could be used for DFI in *A. fumigatus*, although this technique may be restricted to the early steps of the infection process since the sorting of hyphal forms in a FACS may not be feasible. As discussed above, transformation strategies for targeted integration of various constructs including genomic libraries at a preselected chromosomal location are now available in *A. fumigatus*. However, these strategies result in a dramatic reduction of the transformation efficiencies (ca. 2 transformants/µg DNA) and prototrophic transformants obtained in this manner do not always occur by targeted integration [30]. This limits the use of these strategies for the construction of a large collection of *A. fumigatus* strains each carrying an independent gene fusion until new techniques are developed that increase transformation efficiencies and ensure proper targeting of the transforming DNA. In this regard, the construction of collections of *A. fumigatus* strains for use in IVET or DFI might be best achieved with the help of linearized vectors that contain an antibiotic resistance gene for the selection of transformants and a reporter gene located at one end of the vector and that will create random gene fusions upon integration at ectopic sites in the *A. fumigatus* genome.

Conclusion

The establishment of various transformation strategies of *A. fumigatus* during the last 5 years has resulted in the rapid development of the molecular genetics of this organism that will help dissect the structure-function relationships of gene products that are proposed to participate in the cell biology and virulence of the fungus. Furthermore, DNA-mediated transformation enables the development of new strategies to identify those genes that are essential to the infection process or expressed during infection. Improving transformation procedures, stability of autonomously replicating plasmids, and targeting systems, developing insertional mutagenesis techniques with the help of transposable elements, establishing more reliable animal models for invasive aspergillosis or cell culture systems that mimick the early stages of infection will be important tasks for the next years that will contribute to our improved knowledge of the biology of this opportunistic pathogen in the host environment and hopefully unravel novel targets for the development of antifungals.

Acknowledgments

Thanks are due to all the members of the Aspergillus laboratory for fruitful discussions on transformation strategies. Jean-Paul Latgé and August Böck are gratefully acknowledged for their constant support. We thank Jeremy S. Brown, Agnès Aufauvre-Brown and David W. Holden for kindly communicating data prior to publication. The work in the laboratory of A.A. Brakhage is supported by the Deutsche Forschungsgemeinschaft.

References

1 Hensel M, Holden DW: Molecular genetic approaches for the study of virulence in both pathogenic bacteria and fungi. Microbiology 1996;142:1049–1058.

2 Punt PJ, Oliver RP, Dingemanse MA, Pouwels PH, van den Hondel CAMJJ: Transformation of *Aspergillus* based on the Hygromycin B resistance marker from *Escherichia coli*. Gene 1987;56: 117–124.

3 van den Hondel CAMJJ, Punt PJ: Gene transfer systems and vector development for filamentous fungi; in Peberdy JF, Caten CE, Ogden JE, Bennett JE (eds): Applied Molecular Genetics of Fungi. Cambridge, Cambridge University Press, 1991, pp 1–28.

4 Tang CM, Cohen J, Holden DW: An *Aspergillus fumigatus* alkaline protease mutant constructed by gene disruption is deficient in extracellular elastase activity. Mol Microbiol 1992;6:1663–1671.

5 Smith JM, Davies JE, Holden DW: Construction and pathogenicity of *Aspergillus fumigatus* mutants that do not produce the ribotoxin restrictocin. Mol Microbiol 1993;9:1071–1077.

6 Monod M, Paris S, Sarfati J, Jaton-Ogay K, Ave P, Latgé JP: Virulence of alkaline protease-deficient mutants of *Aspergillus fumigatus*. FEMS Microbiol Lett 1993;106:39–46.

7 Paris S, Monod M, Diaquin M, Lamy B, Arruda LK, Punt PJ, Latgé JP: A transformant *Aspergillus fumigatus* deficient in the antigenic cytotoxin ASPFI. FEMS Microbiol Lett 1993;111:31–36.

8 Smith JM, Tang CM, Noorden SV, Holden DW: Virulence of *Aspergillus fumigatus* double mutants lacking restrictocin and an alkaline protease in a low-dose model of invasive pulmonary aspergillosis. Infect Immun 1994;62:5247–5254.

9 Jaton-Ogay K, Paris S, Huerre M, Quadroni M, Falchetto R, Togni G, Latgé JP, Monod M: Cloning and disruption of the gene encoding an extracellular metalloprotease of *Aspergillus fumigatus*. Mol Microbiol 1994;14:917–928.

10 Ibrahim-Granet O, d'Enfert C: The *Aspergillus fumigatus mepB* gene encodes an 82 kDa intracellular metalloproteinase structurally related to mammalian thimet oligopeptidases. Microbiology 1997; 143:2247–2253.

11 Reichard U, Monod M, Odds F, Ruchel R: Virulence of an aspergillopepsin-deficient mutant of *Aspergillus fumigatus* and evidence for another aspartic proteinase linked to the fungal cell wall. J Med Vet Mycol 1997;35:189–196.

12 Mellado E, Aufauvre-Brown A, Gow NAR, Holden DW: The *Aspergillus fumigatus chsC* and *chsG* genes encode class III chitin synthases with different functions. Mol Microbiol 1996;20:667–679.

13 Mellado E, Specht CA, Robbins PW, Holden DW: Cloning and characterization of *chsD*, a chitin synthase-like gene of *Aspergillus fumigatus*. FEMS Microbiol Lett 1996;143:69–76.

14 Aufauvre-Brown A, Mellado E, Gow NA, Holden DW: *Aspergillus fumigatus chsE*: A gene related to *CHS3* of *Saccharomyces cerevisiae* and important for hyphal growth and conidiophore development but not pathogenicity. Fungal Genet Biol 1996;21:141–152.

15 Tsai HF, Washburn RG, Chang YC, Kwonchung KJ: *Aspergillus fumigatus arp1* modulates conidial pigmentation and complement deposition. Mol Microbiol 1997;26:175–183.

16 Thau N, Monod M, Crestani B, Rolland C, Tronchin G, Latgé JP, Paris S: *Rodletless* mutants of *Aspergillus fumigatus*. Infect Immun 1994;62:4380–4388.

17 Calera J-A, Paris S, Monod M, Hamilton AJ, Debeaupuis JP, Diaquin M, Lopez-Medrano R, Leal F, Latgé J-P: Cloning and disruption of the antigenic catalase gene of *Aspergillus fumigatus*. Infect Immun 1997;65:4718–4724.

18 May GS, Gambino J, Weatherbee JA, Morris NR: Identification and functional analysis of beta-tubulin genes by site-specific integrative transformation in *Aspergillus nidulans*. J Cell Biol 1985; 101:712–719.

19 Orbach MJ, Porro EB, Yanofsky C: Cloning and characterization of the gene for beta-tubulin from a benomyl-resistant mutant of *Neurospora crassa* and its use as a dominant selectable marker. Mol Cell Biol 1986;6:2452–2461.

20 Seip ER, Woloshuk CP, Payne GA, Curtis SE: Isolation and sequence analysis of a beta-tubulin gene from *Aspergillus flavus* and its use as a selectable marker. Appl Environ Microbiol 1990;56: 3686–3692.

21 Avalos J, Geever RF, Case ME: Bialaphos resistance as a dominant selectable marker in *Neurospora crassa*. Curr Genet 1989;16:369–372.
22 Straubinger B, Straubinger E, Wirsel S, Turgeon G, Yoder O: Versatile fungal transformation vectors carrying the selectable *bar* gene of *Streptomyces hygroscopicus*. Fungal Genet Newslett 1992;39: 82–83.
23 Rothstein R: Targeting, disruption, replacement and allele rescue: Integrative DNA transformation in yeast. Methods Enzymol 1991;194:291–301.
24 Sikorski RS, Boeke JD: In vitro mutagenesis and plasmid shuffling: From cloned gene to mutant yeast. Methods Enzymol 1991;194:302–318.
25 Timberlake WE: Cloning and analysis of fungal genes; in Bennett JW, Lasure LL (eds): More Gene Manipulations in Fungi. Oxford, Academic Press, 1991, pp 51–85.
26 Boeke JD, Lacroute F, Fink GR: A positive selection for mutants lacking orotidine-5′-phosphate decarboxylase activity in yeast: 5-Fluoro-orotic acid resistance. Mol Gen Genet 1984;197:345–346.
27 d'Enfert C: Selection of multiple disruption events in *Aspergillus fumigatus* using the orotidine-5′-decarboxylase gene, *pyrG* as a unique transformation marker. Curr Genet 1996;30:76–82.
28 Kinghorn J: Genetic, biochemical, and structural organization of *Aspergillus nidulans crnA-niiA-niaD* gene cluster; in Wray J, Kinghorn J (eds): Molecular Genetics Aspects of Nitrate Assimilation. Oxford, Oxford Science Publishers, 1989, pp 69–87.
29 van Hartingsveldt W, Mattern IE, van Zeijl CMJ, Pouwels PH, van den Hondel CAMJJ: Development of a homologous transformation system for *Aspergillus niger* based on the *pyrG* gene. Mol Gen Genet 1987;206:71–75.
30 Weidner G, d'Enfert C, Koch A, Mol PC, Brakhage AA: Development of a homologous transformation system for the pathogenic fungus *Aspergillus fumigatus* based on the *pyrG* gene encoding orotidine monophosphate decarboxylase. Current Genet 1998;33:378–385.
31 Gouka RJ, Hessing JG, Stam H, Musters W, van den Hondel CAMJJ: A novel strategy for the isolation of defined *pyrG* mutants and the development of a site-specific integration system for *Aspergillus awamori*. Current Genet 1995;27:536–540.
32 Unkles SE, Campbell EI, Carrez D, Grieve C, Contreras R, Fiers W, van den Hondel CAMJJ, Kinghorn JR: Transformation *Aspergillus niger* with the homologous nitrate reductase gene. Gene 1989;78:157–166.
33 Borgia PT, Dodge CL, Eagleton LE, Adams TH: Bidirectional gene transfer between *Aspergillus fumigatus* and *Aspergillus nidulans*. FEMS Microbiol Lett 1994;122:227–232.
34 Gavrias V, Thiede N, Iartchouk N, Saxton J-A, Hitchcock C, Timberlake WE, Koltin Y: The construction of vectors and strains for efficient gene disruption in *Aspergillus fumigatus*. Fungal Genet Newslett 1997;44A:24.
35 d'Enfert C, Diaquin M, Delit A, Wuscher N, Debeaupuis JP, Huerre M, Latgé JP: Attenuated virulence of uridine-uracil auxotrophs of *Aspergillus fumigatus*. Infect Immun 1996;64:4401–4405.
36 Tang CM, Smith JM, Arst HN, Holden DW: Virulence studies of *Aspergillus nidulans* mutants requiring lysine or *p*-aminobenzoic acid in invasive pulmonary aspergillosis. Infect Immun 1994; 62:5255–5260.
37 Brown JS, Aufauvre-Brown A, Tiffin N, Holden DW: Identification of virulence determinants of *A. fumigatus* using a REMI/STM approach. Fungal Genet Newslett 1997;44A:27.
38 Peberdy JF: Fungi without coats – Protoplasts as tools for mycological research. Mycol Res 1989; 93:1–20.
39 Paris S: Isolation of protease negative mutants of *Aspergillus fumigatus* by insertion of a disrupted gene; in Maresca B, Kobayashi GS (eds): Molecular Biology of Pathogenic Fungi: A Laboratory Manual. New York, Telos Press, 1994, pp 49–55.
40 Smith JM: Construction of mutants of *Aspergillus fumigatus* by gene disruption; in Maresca B, Kobayashi GS (eds): Molecular Biology of Pathogenic Fungi: A Laboratory Manual. New York, Telos Press, 1994, pp 41–47.
41 Sanchez O, Aguirre J: Efficient transformation of *Aspergillus nidulans* by electroporation of germinated conidia. Fungal Genet Newslett 1996;43:48–51.
42 Aleksenko A, Clutterbuck AJ: Autonomous plasmid replication in *Aspergillus nidulans*: AMA1 and MATE elements. Fungal Genet Biol 1997;21:373–387.

43 Berg C, Garber ED: A genetic analysis of color mutants of *Aspergillus fumigatus.* Genetics 1962; 47:1139–1146.
44 Schiestl RH, Petes TD: Integration of DNA fragments by illegitimate recombination in *Saccharomyces cerevisiae.* Proc Natl Acad Sci USA 1991;88:7585–7589.
45 Shi Z, Christian D, Leung H: Enhanced transformation in *Magnaporthe grisea* by restriction enzyme mediated integration of plasmid DNA. Phytopathology 1995;85:329–333.
46 Lu SW, Lyngholm L, Yang G, Bronson C, Yoder OC, Turgeon BG: Tagged mutations at the *Tox1* locus of *Cochliobolus heterostrophus* by restriction enzyme-mediated integration. Proc Natl Acad Sci USA 1994;91:12649–12653.
47 Bölker M, Böhnert HU, Braun KH, Görl J, Kahmann R: Tagging pathogenicity genes in *Ustilago maydis* by restriction enzyme-mediated integration (REMI). Mol Gen Genet 1995;248:547–552.
48 Aufauvre-Brown A, Tang CM, Holden DW: Detection of gene-disruption events in *Aspergillus* transformants by polymerase chain reaction direct from conidiospores. Curr Genet 1993;24:177–178.
49 d'Enfert C, Fontaine T: Molecular characterization of the *Aspergillus nidulans treA* gene encoding an acid trehalase required for growth on trehalose. Mol Microbiol 1997;24:203–216.
50 Mellado E, Aufauvre-Brown A, Specht CA, Robbins PW, Holden DW: A multigene family related to chitin synthase genes of yeast in the opportunistic pathogen *Aspergillus fumigatus.* Mol Gen Genet 1995;246:353–359.
51 Jaton-Ogay K, Suter M, Crameri R, Falchetto R, Fatih A, Monod M: Nucleotide sequence of a genomic and a cDNA clone encoding an extracellular alkaline protease of *Aspergillus fumigatus.* FEMS Microbiol Lett 1992;92:163–168.
52 Ramesh MV, Sirakova TD, Kolattukudy PE: Cloning and characterization of the cDNAs and genes (*mep20*) encoding homologous metalloproteinases from *Aspergillus flavus* and *Aspergillus fumigatus.* Gene 1995;165:121–125.
53 Beauvais A, Monod M, Debeaupuis JP, Diaquin M, Kobayashi H, Latgé JP: Biochemical and antigenic characterization of a new dipeptidyl-peptidase isolated from *Aspergillus fumigatus.* J Biol Chem 1997;272:6238–6244.
54 Beauvais A, Monod M, Wyniger J, Debeaupuis JP, Grouzmann E, Brakch N, Svab J, Hovanessian AG, Latgé JP: Dipeptidyl-peptidase IV secreted by *Aspergillus fumigatus*, a fungus pathogenic to humans. Infect Immun 1997;65:3042–3047.
55 Osmani SA, Engle DB, Doonan JH, Morris NR: Spindle formation and chromatin condensation in cells blocked at interphase by mutation of a negative cell cycle control gene. Cell 1988;52:241–251.
56 Suelmann R, Sievers N, Fischer R: Nuclear traffic in fungal hyphae: In vivo study of nuclear migration and positioning in *Aspergillus nidulans.* Mol Microbiol 1997;25:757–769.
57 Som T, Kolaparthi VSR: Developmental decisions in *Aspergillus nidulans* are modulated by ras activity. Mol Cell Biol 1994;14:5333–5348.
58 Valdivia RH, Falkow S: Probing bacterial gene expression within host cells. Trends Microbiol 1997; 5:360–363.
59 Heithoff DM, Conner CP, Mahan MJ: Dissecting the biology of a pathogen during infection. Trends Microbiol 1997;5:509–513.
60 Hensel M, Shea JE, Gleeson C, Jones MD, Dalton E, Holden DW: Simultaneous identification of bacterial virulence genes by negative selection. Science 1995;269:400–403.
61 Mahan MJ, Slauch JM, Mekalanos JJ: Selection of bacterial virulence genes that are specifically induced in host tissues. Science 1993;259:686–688.
62 Valvidia RH, Falkow S: Fluorescence-based isolation of bacterial genes expressed within host cells. Science 1997;277:2007–2011.

Christophe d'Enfert, Laboratoire des Aspergillus, Institut Pasteur,
25, rue Dr. Roux, F–75724 Paris Cedex 15 (France)
Tel. +33 1 40 61 32 57, Fax +33 1 45 68 87 90, E-Mail denfert@pasteur.fr

Brakhage AA, Jahn B, Schmidt A (eds): *Aspergillus fumigatus.*
Contrib Microbiol. Basel, Karger, 1999, vol 2, pp 167–181

Interactions between *Aspergillus fumigatus* and Host Matrix Proteins

Jean-Philippe Bouchara, Myriam Sanchez, Karine Esnault, Guy Tronchin

Groupe d'Etude des 'Interactions Hôte-Parasite', Laboratoire de Parasitologie-Mycologie, Centre Hospitalier Universitaire, Angers, France

Aspergillus fumigatus, the main causative agent of human aspergillosis, is usually a saprophyte, growing and sporulating in a humid environment on decaying organic matter. Therefore, *A. fumigatus* is an opportunistically pathogenic fungus eliciting by inhalation of its airborne conidia a wide spectrum of respiratory diseases ranging from localized infections such as sinusitis or bronchopulmonary aspergilloma to invasive, rapidly progressing and often fatal infections [1, 2]. Because of the increasing prevalence of these mycoses, along with the increasing number of immunocompromised patients, of their high mortality rate and their social cost, elucidation of the virulence determinants of the fungus allowing its installation within the host and the initiation of disease has become an important area of research over the past decade [3].

Adherence of microorganisms to the host epithelial barriers constitutes a critical step in the development of infections since it conditions the expression of their pathogenic potential and therefore is an attractive target for the design of new antimicrobial agents such as vaccines or adhesion blockers. Adherence is often the result of specific interactions (inhibitable by macromolecules or by their peptide or sugar moieties) between well-localized surface components of the interacting elements. These interactions require a close approach and a stereochemical complementarity between surface components of the microorganism, designated adhesins or receptors, and some host adhesive proteins called ligands. Over the past decade, it has become increasingly obvious that extracellular matrix (ECM) proteins or glycoproteins, particularly fibronectin and laminin, may act as mediators for the attachment of colonizing microorganisms. Many microorganisms express cell surface adhesins that recognize the ECM components and Patti et al. [4] have proposed the acronym

MSCRAMMs (microbial surface components recognizing the adhesive matrix molecules) to designate this class of adhesins. In *A. fumigatus*, our laboratory and others have focused their research on the interactions of the fungus with fibrinogen and laminin whose role in the adherence phenomena is largely documented. However, specific interactions with fibronectin and collagen, as well as the role of hydrophobic interactions in adherence of *A. fumigatus* will also be briefly discussed.

Structure and Functions of Fibrinogen and Basement Membrane Laminin

With a concentration of 2–4 g/l under physiological conditions, fibrinogen is the major plasma glycoprotein [5]. This dimeric glycoprotein of 340 kD comprises two sets of three nonidentical polypeptide chains, designated Aα, Bβ and γ, held together by numerous disulfide bonds. The native molecule presents a polydomainal structure with two identical globular domains, the D domains, at the extremities of the molecule and a central globular domain, the E domain, linked by interdomainal supercoiled α-helices. Fibrinogen, which is synthesized by hepatocytes, plays important roles in the coagulation pathway and wound healing and its synthesis is stimulated in inflammation due to mechanical or chemical tissue injury and by infections. In addition, recent in vitro experiments have shown that lung epithelial cells also synthesize fibrinogen and secrete it in the basolateral direction, suggesting that it is also a matrix protein [6]. Therefore, for numerous microorganisms, the recognition of fibrinogen may constitute a mechanism of attachment to the host tissues. Thus, several bacteria or fungi express specific surface receptors for fibrinogen which could allow them to attach to damaged epithelia or endothelia, but also to blood platelets and fibrin-platelet clots [4].

Laminin is an ECM protein found exclusively in basement membranes which underlie epithelia and endothelia. This major structural component of basement membranes is a large glycoprotein of 900 kD composed of three nonidentical polypeptide chains (A, B_1 and B_2) assembled by disulfide bonds in a cross-shaped structure [7]. Through its self-assembly and its interactions with the other components of basement membranes, namely type IV collagen, entactin/nidogen, heparan sulfate proteoglycans and osteonectin, laminin plays a key role in the assembly of basement membranes. This multifunctional glycoprotein also promotes growth, spreading, migration, proliferation and differentiation of cells and supports cell adhesion. Hence cell-matrix interactions are of crucial importance in a variety of physiological or pathological events including cellular cohesion inside tissues and diapedesis, but also tumor invasion and metastasis. Specific laminin receptors are described on cells that normally interact with basement membranes, for example epithelial and endothelial cells, as well as on cells that extravasate such as neutrophils or metastatic tumor cells.

Some of these laminin-binding proteins belong to the integrin family and recognize the tripeptide sequence RGD (single-letter code for amino acids) localized on the short arm of the A chain in laminin P_1, a fragment which results from pepsin digestion and corresponds to the central region of the short arms of the cross. Other integrins bind to the peptide sequence IKVAV localized on the long arm of the A chain in the E_8 fragment obtained by mild elastase digestion. Non-integrin receptors are also described, mainly a monomeric 67- to 70-kD receptor which binds laminin with a high affinity and recognizes the peptide sequence YIGSR localized on the B_1 chain, in the fragment P_1, and the 67-kD elastin/laminin receptor which binds laminin through the peptide LGTIPG also localized on the B_1 chain.

Various bacteria as well as some protozoa also recognize laminin through specific surface receptors. Likewise, fungal infections may be initiated by the attachment of the pathogen to basement membrane laminin. Thus receptors for laminin are described on the human pathogenic fungi *Candida albicans* [8], *Rhizopus oryzae* [9], *Histoplasma capsulatum* [10] and *Paracoccidioides brasiliensis* [11], but also on the entomopathogenic fungi *Paecilomyces farinosus* [12] and *Nomuraea rileyi* [13] and some fish pathogenic yeasts *Candida zeylanoides, Rhodotorula rubra* and *Rhodotorula glutinis* [14].

Possible Role of Matrix Proteins in the Host Tissue Colonization

Several studies have demonstrated the interactions between *A. fumigatus* conidia and some matrix proteins such as fibrinogen and laminin, and suggested a role for these interactions in the host tissue adherence. In healthy individuals, ECM components are normally unexposed and it has been suggested that a direct attachment of the inhaled conidia to the alveolar epithelial cells occurs via some pneumocyte membrane glycosphingolipids, particularly lactosylceramide, yet adherence may also take place at sites of intercellular tight junctions [15].

However, epithelial tissue damage often accompanies the predisposing factors for aspergillosis. In bronchopulmonary aspergilloma, the development of the fungus as a fungus ball arises above all by colonization of preexisting cavities, mainly a tuberculous cavitation [1]. Likewise, among the adverse effects of chemotherapy, which is the main risk factor for invasive pulmonary aspergillosis in bone marrow transplant recipients [2, 16], is epithelial tissue injury.

Despite the impairment of the mucociliary clearance, and the entrapment of the airborne conidia in the viscous mucosal secretions due to the defective chloride channel, the fungal colonization of the airways in patients with cystic fibrosis (CF) is very rare in young children and it usually follows bacterial infections. Studying the epidemiology of fungal infections in CF patients, we found that the mean age at date of first colonization of the airways by *A. fumigatus* was 12.3 years, whereas it was 5.4 and 8.1 for *Staphylococcus aureus* and *Pseudomonas aeruginosa*, respectively [17]. Thus previous epithelial lesions consecutive to chronic bacterial infections and induced by the bacterial proteinases or by the neutrophil elastase released by local activated leukocytes [18] are required for the establishment of fungi in the respiratory tract of CF patients. Similarly, with the prolonged survival of AIDS patients, there have been in recent years increasing reports of pulmonary aspergillosis [19]. Although inconclusive, the high frequency of preceding or intercurrent bronchopulmonary infections due to cytomegalovirus, *Mycobacterium avium* complex or *Pneumocystis carinii* suggests that they may be a predisposing factor.

Alternatively, it has been suggested recently using A549 and NCI-H292 cultured cells (two human epithelial cell lines with characteristics of pneumocytes type II and bronchial epithelial cells, respectively) that the subtilisin-like serine protease secreted by *A. fumigatus* may lead to the detachment of epithelial cells and the release of proinflammatory cytokines [20]. These cellular lesions – preexisting or induced by the fungus – lead to an inflammatory reaction and to the exposure of the subepithelial basement membrane and the underlying interstitium, making accessible laminin, fibronectin, and type I and IV collagens. Therefore, it was tempting to speculate that ECM components and the fibrinogen deposits formed at the surface of wounded epithelia in response to this inflammatory reaction could mediate the adherence of the conidia which are the infecting form of the fungus.

Evidence for These Interactions and Their Role in Adherence

Interaction with fibrinogen was first evidenced using ^{125}I-labeled human fibrinogen since we showed that, among about thirty filamentous fungi belonging to different groups (opportunistic fungi, dermatophytes, strictly saprophyte or phytopathogenic fungi), only some pathogenic aspergilli bound fibrinogen [21]. The binding of fibrinogen and laminin, visualized by immunofluorescence, appears to be restricted to conidia and mother cells of germ tubes [22–24]. Conversely in *C. albicans*, both glycoproteins bind exclusively to the hyphal surface [25] and these opposite binding patterns may reflect the differences in the modes of infestation between candidiasis and aspergillosis.

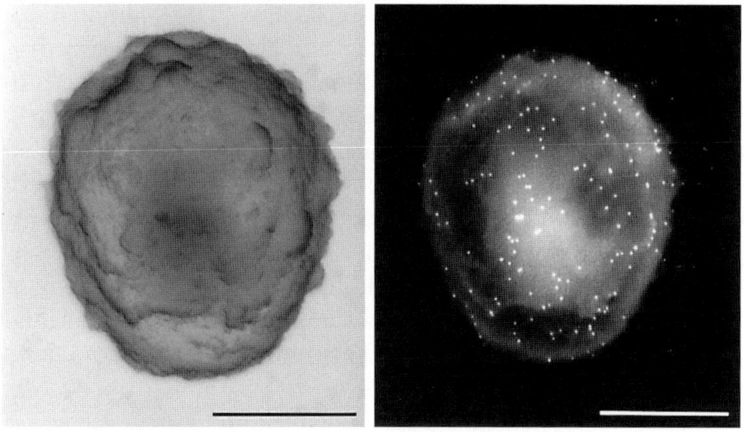

Fig. 1. Scanning immunoelectron micrographs (combined secondary electron signal on the left, and backscattered electron image of the same cell on the right) of resting conidia labeled with laminin, anti-laminin antibodies, and protein A-gold (20 nm). The labeling is associated with the protrusions of the cell wall. Bars = 1 μm.

Electron microscopy performed with cells at different stages of germination, from resting conidia to germ tubes, confirmed this binding pattern and permitted precise localization of the binding sites [22, 23]. Incubation of the cells with fibrinogen or laminin, followed by a rabbit immune serum and protein A-gold resulted in an intense labeling at the surface of resting conidia. Labeling was associated with the protrusions of their cell wall (fig. 1), a localization which was also seen for fibronectin binding by immunofluorescence confocal microscopy [26]. Moreover, swelling and germination, which lead to the thinning and disintegration of the outer cell wall layer, resulted in a decrease in the intensity of the labeling which became very scattered over the surface of the swollen conidia and of the mother cells of germ tubes.

Adherence assays to purified plasma or ECM components immobilized onto microtiter plates were developed in our laboratory. In these assays, the adherent conidia were counted by phase-contrast microscopy or detected using specific antibodies [23, 27, 28]. Thus, we demonstrated that conidia adhered avidly to laminin or collagen substrates, but not to the glycosaminoglycans, heparan sulfate, keratan sulfate and chondroitin sulfates A and B. Fibrinogen also seemed to be a good candidate for mediating the adherence. In contrast, a very weak or no adherence was observed for wells coated with human or bovine serum albumin as well as with human fibronectin or immunoglobulins G. Nevertheless, recent experiments performed by others [26, 29, 30] suggest fibronectin as a ligand for *A. fumigatus* conidia, although they attach to fibronectin less efficiently than to fibrinogen or laminin [29]. Additional evidence for the role of these interactions in the host tissue adherence derives from the ability of soluble fibrinogen [29], laminin [31] or fibronectin [31] to inhibit the attachment of the conidia to A549 cells.

Adherence assays also showed the specificity of the attachment of conidia to immobilized fibrinogen, laminin, fibronectin or collagen. Preincubation of the cells with the soluble ligand, or with polyclonal antibodies directed towards these glycoproteins, inhibited their attachment to the respective ligand, demonstrating that the interactions are specific [23, 26, 28–30]. Moreover, study of the adherence of numerous strains from human, animal or environmental origins to various mammalian fibrinogens did not reveal any particular affinity of strains for some animal species [28].

The binding sites for *A. fumigatus* conidia on the fibrinogen molecule are localized in the outer D domains [28]. Soluble fragment D, but not fragment E, both obtained by plasmin digestion of the fibrinogen molecule, markedly reduced the adherence to immobilized fibrinogen. Likewise, conidia adhered avidly to immobilized fragment D, but not to immobilized fragment E. These results highlight the role of this interaction in the pathogenesis of the fungus, since in the native molecule only the D domains are accessible. In addition, these assays allowed us to localize the binding sites on the laminin molecule [23]. The binding sites for *A. fumigatus* conidia appeared to be located in laminin P_1. Conidia adhered avidly to immobilized P_1, whereas no attachment could be observed with immobilized E_8.

Flow cytometry using fluorescein isothiocyanate (FITC)-conjugated fibrinogen, as well as binding assays performed with radiolabeled human fibrinogen or its fragments D and E, confirmed the involvement of the D domains, and the saturability and specificity of the binding sites at the conidial surface [22, 28]. Incubation of resting conidia with FITC-fibrinogen and a 10-fold molar excess of unlabeled fibrinogen or fragment D resulted in a marked decrease in the fluorescence intensity of the cells, compared to that obtained in the absence of any potential inhibitor [28]. In contrast, no inhibition was observed in the presence of equimolar concentrations of fragment E or BSA. Likewise, after incubation of the conidia with iodinated purified fibrinogen or its fragments D and E, equimolar concentrations of fibrinogen and fragment D were recovered at the conidial surface, whereas a very weak amount of fragment E bound to the fungal elements under the same conditions [22].

The binding sites for fibrinogen fragment D at the conidial surface were found to be heat- and trypsin-sensitive, indicating a proteinaceous nature, and showed characteristics of true receptors [22]. Incubation of equal numbers of conidia with increasing amounts of radiolabeled fragment D demonstrated the saturability of the binding. Furthermore, competition experiments performed by incubating the cells in the presence of increasing concentrations of radiolabeled fragment D and a 100-fold molar excess of unlabeled ligand allowed us to calculate the specific binding. Analysis of the data obtained using the Scatchard equation revealed an average number of receptors of about 1,200 per conidium, and a single dissociation constant of approximately 10^{-7} *M*.

Multiplicity of the Adherence Systems in *A. fumigatus*

Sialic Acid-Dependent Recognition of Laminin and Fibrinogen

Recent ligand-blotting experiments enabled the identification of the laminin receptors from a conidial cell wall extract obtained by heat treatment of the cells in the presence of sodium dodecyl sulfate (SDS) and dithiothreitol [24]. After SDS-PAGE and Western-blotting, followed by incubation of the blot with laminin, rabbit anti-laminin immune serum, and peroxidase-conjugated anti-rabbit IgG antibodies, only one band was detected with a molecular mass of 72 kD, and therefore close to that of other fungal receptors for laminin. In addition to the 43-kD glycoprotein, a second laminin-binding protein of 72–75 kD has been reported in *Paracoccidioides brasiliensis* [Gesztesi, unpubl. data].

Candida albicans germ tubes also express a 68-kD component which binds laminin with a high affinity [8]. Although it remains questionable, it has been suggested that this 68-kD mannoprotein cross-reacts with the human 67-kD high-affinity laminin receptor (67-HLR) [32] and recognizes laminin by its peptide sequence CDPGYIGSR [33] as it occurs for the 67-HLR. Strikingly, similar ligand-blotting experiments performed by Gill et al. [26] on *A. fumigatus* conidial extracts allowed the identification of the laminin receptor as a 37-kD component, a molecular mass similar to that of the laminin receptor identified by the same group on *C. albicans* [32] and of the mammalian 37-kD protein which is thought to be a precursor of the 67-HLR [34].

The fibrinogen receptors in *A. fumigatus* have not yet been identified, but there is now accumulated evidence indicating that the same fungal components are involved in the binding of both fibrinogen and laminin. In addition to the similar binding pattern that we previously mentioned for these two host proteins, competition experiments showed a cross-reactivity between laminin and fibrinogen. The binding of FITC-fibrinogen to the conidia and the attachment of conidia to immobilized fibrinogen were markedly reduced in the presence of laminin [28]. Moreover, fibrinogen inhibited the adherence of the conidia to immobilized laminin [23] as well as the binding of FITC-conjugated laminin [35]. Synthetic peptides which mimic adhesive peptide sequences common to both ligands, i.e. RGD-containing peptides (RGD, RGDS, GRDS and GPRP), as well as peptide sequences specifically found in laminin (YIGSR and

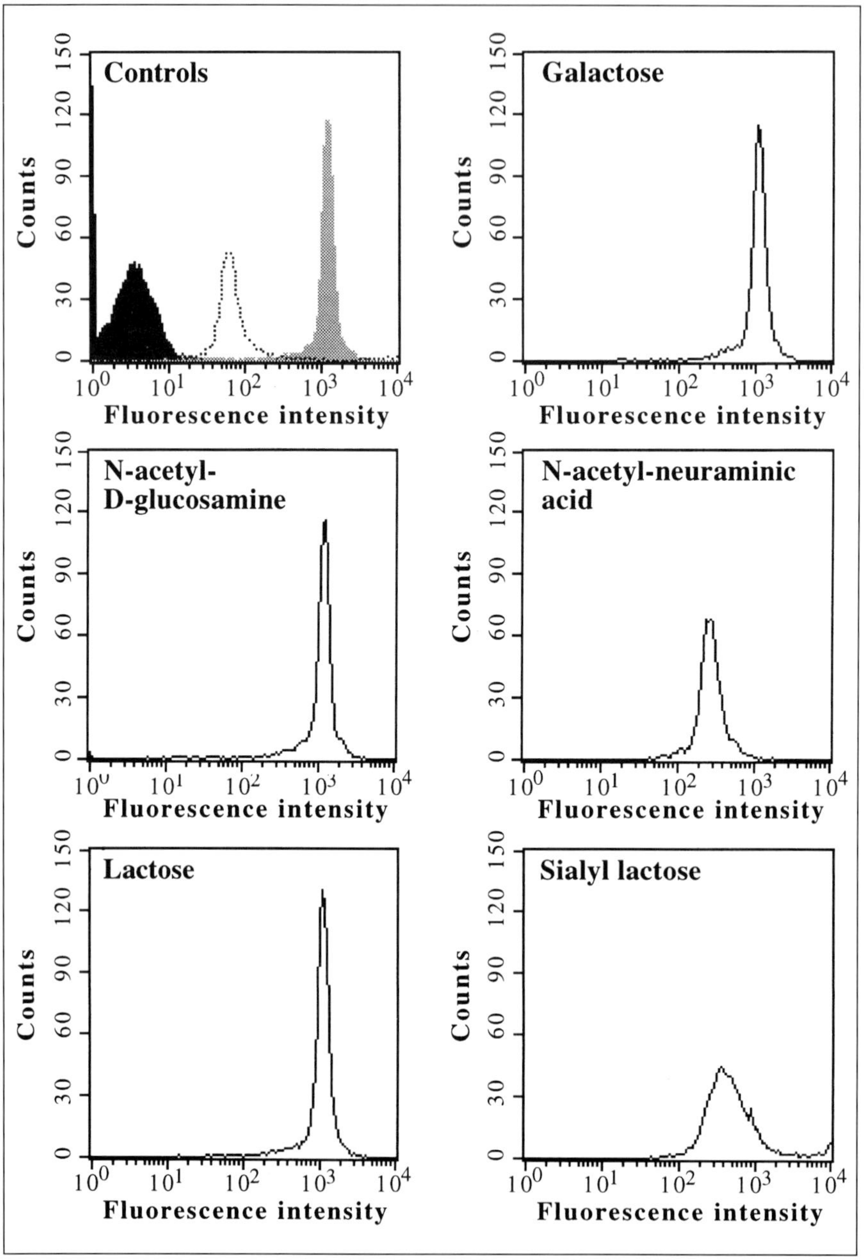
Controls
Galactose
N-acetyl-D-glucosamine
N-acetyl-neuraminic acid
Lactose
Sialyl lactose
Counts
Fluorescence intensity

CDPGYIGSR or their amide forms, IKVAV and LGTIPG) or in fibrinogen (fibrinogen γ-chain 12-mer peptide) failed to inhibit the binding of FITC-labeled fibrinogen and laminin [28, 35]. These results, added to the aggregation of conidia incubated with laminin which suggests the presence of multiple binding sites on the laminin molecule [Tronchin, unpubl. data], led us to consider the carbohydrates as the recognition motifs. Laminin and fibrinogen Bβ- and γ-chains present closely related N-linked carbohydrate chains composed of N-acetyl lactosamine or polylactosamine side chains with a terminal sialic acid residue, branched by mannose residues on the glycosidic core. Among the carbohydrates tested constitutive of the fibrinogen or laminin sugar chains or related to them, only N-acetyl-neuraminic acid and sialyl lactose inhibited the binding of both glycoproteins (fig. 2) [35].

RGD-Dependent Recognition of Fibronectin

Although it has been reported that fibronectin has a binding pattern similar to that of laminin [26, 30] and a cross-reactivity with laminin [26, 30], it seems obvious now that receptors distinct from those for laminin and fibrinogen mediate the binding of fibronectin. Attachment of the conidia to immobilized fibronectin was inhibited by RGD-containing peptides, but not the attachment to immobilized laminin [26, 29] or fibrinogen [29] which confirms the lack of inhibition potency of these synthetic peptides in our hands. In addition, the fibronectin receptors identified by ligand-blotting experiments from a whole conidial homogenate or a 2-mercaptoethanol extract of isolated conidial walls were represented by two polypeptides of 23 and 30 kD [30]. These molecular masses are different from those reported by the same group and ours for the laminin receptor.

Other Adhesive Systems

As previously mentioned, type I collagen from interstitium tissue and basement membrane type IV collagen may also serve as ligands for the attachment of conidia. Despite the relatively few data available, it seems likely that the recognition of collagen, especially type IV collagen, is mediated by mechanisms

Fig. 2. Influence of carbohydrates on the binding of laminin to *A. fumigatus* conidia. Conidia were incubated with FITC-laminin in the presence of different carbohydrates at 200 mM final concentration (except for sialyl lactose 100 mM). After 30 min at 37 °C, the fluorescence at the cell surface was quantified by flow cytometry. x axis = Fluorescence intensity expressed in arbitrary units on a logarithmic scale; y axis = number of fluorescent cells. Controls correspond to cells incubated in the buffer alone (solid black curve showing the absence of autofluorescence), with FITC-laminin alone (total binding; solid gray curve) or with FITC-laminin and a 10-fold molar excess of unlabeled ligand (nonspecific binding; dotted line).

distinct from those involved in fibronectin binding. Adherence of conidia to immobilized type I and IV collagen was markedly reduced in the presence of soluble laminin, but only weakly in the presence of fibronectin [26]. Moreover, the binding of FITC-conjugated type I and IV collagen was not affected by RGD peptides [26], nor the attachment of conidia to immobilized type IV collagen [26, 29], whereas conflicting results were reported for their attachment to type I collagen [26, 29].

A 32-kD fucose-specific lectin, concentrated in conidia rather than in mycelia, has been identified recently in *A. fumigatus* [36]. Further studies are needed to confirm the presence of this lectin which could provide a mechanism for the attachment of conidia to fucosylated antigens of the epithelial cell membrane, but also for the recognition of matrix proteins through the fucose residues of their carbohydrate side chains. We did not find by flow cytometry any inhibition of the binding of FITC-laminin, fibrinogen or fibronectin to the conidia in the presence of *L*-fucose [Sanchez, unpubl. data]. Likewise, flow cytometry using FITC-conjugated neoglycoproteins (glucosyl, mannosyl, lactosyl or fucosyl-BSA) did not reveal a particular enhancement of the binding by the fucose residues as compared with the fixation of other neoglycoproteins or BSA. Although all the strains that we have studied presented a similar adhesive behaviour, strain variation or differences in culture conditions may account for these conflicting results.

Influence of Culture Conditions on the Adhesive Properties of the Conidia

As we have seen, investigations on the adherence mechanisms in *A. fumigatus* have been focused primarily on specific interactions. Long-range nonspecific interactions, due to the overall surface properties of the cells such as charge or hydrophobicity, are involved in bacterial adherence, and therefore could mediate the attachment of the inhaled conidia to the lung tissues. Moreover, electrostatic and hydrophobic forces may also be involved in the attachment to a particular ligand. For example, Silva et al. [37] showed a greater adherence of hydrophobic strains of *C. albicans* to immobilized fibrinogen and ECM components, compared to hydrophilic strains.

In *A. fumigatus*, a gene designated *rodA* [38] or *HYP1* [39] encoding a moderatly hydrophobic protein of 13.4 kD which presents 85% similarity to the hydrophobin RODA of *A. nidulans* [40], has been described recently. However, the disruption of this gene did not affect the adherence of conidia to rat pneumocytes, as well as their attachment to immobilized laminin or fibrinogen, suggesting that this hydrophobin does not play a critical role in adherence

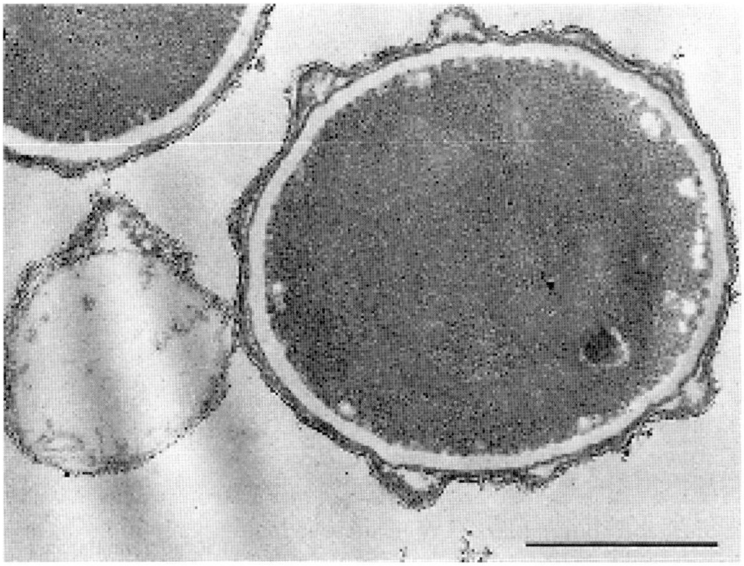

Fig. 3. Ultrathin section of *A. fumigatus* conidia from YPDA culture medium labeled with cationized ferritin. An intense labeling is detected over the whole surface of the conidia. Bar = 1 μ*m*.

[38]. More recently, a second gene encoding another hydrophobin, *dewA*, has been described in *A. nidulans* [41] and it seems likely that several hydrophobic proteins also exist in *A. fumigatus*. Therefore, in an attempt to define the relationships between the electrostatic charge of the conidia, the cell surface hydrophobicity and their ability to bind the host matrix proteins, we analyzed the effects of culture conditions since they greatly influence the composition of the cell wall and subsequently the surface properties for several bacteria and fungi [14, 42, 43].

Resting conidia of *A. fumigatus* are hydrophobic and present an electronegative charge as visualized by electron microscopy using cationized ferritin (fig. 3). However, these surface properties are dependent on the growth medium [Bouchara, unpubl. data]. Cultivation of the fungus on Czapek agar instead of yeast extract-peptone-dextrose-agar (YPDA) resulted in a diminished cell surface hydrophobicity, as revealed by two-phase partitioning and salt aggregation test, and in a parallel decrease in the ability of the conidia to bind FITC-labeled laminin and fibrinogen (table 1). A slight increase in the electronegative charge of the conidia measured by microelectrophoresis was also observed, associated with a complete loss of their ability to bind fibronectin. These results confirm, if needed, the idea that the fibronectin receptors are distinct

Table 1. Influence of culture conditions on the surface properties of *A. fumigatus* conidia

	Expression or affinity of fungal receptors	Electronegative charge	Cell surface hydrophobicity
Culture medium			
Czapek vs. YPDA medium	↘ (complete loss for fibronectin binding)	↗	↘
Glucose or galactose vs. saccharose	no change	↗	no change
Incubation time (24 h to 9 days)	↗ maximum at days 5–7	no change	↗ maximum at day 5

from those for laminin and fibrinogen. Additionally, the use of glucose or galactose instead of saccharose as the carbon source in Czapek agar resulted in a marked increase in the electronegative charge of the conidia, whereas their ability to bind laminin, fibrinogen or fibronectin was not modified, nor their cell surface hydrophobicity.

Incubation time also influences the surface properties [Bouchara, unpubl. data]. Flow cytometric analysis of the fluorescence of conidia from 24-hour to 9-day-old cultures on YPDA incubated with FITC-conjugated laminin, fibrinogen or fibronectin revealed a similar evolution for the total binding of the three ligands which reached a maximum for 5- to 7-day-old cultures whereas the nonspecific binding remained unchanged (table 1). This confirms previous observations suggesting that maturation of the conidia correlates with an increased density, or affinity, of the receptor sites at the conidial surface [28, 35]. The cell surface hydrophobicity also increased with the age of the conidia. In contrast, the zeta potential assessed in parallel remained unchanged.

In summary, adherence of *A. fumigatus* conidia to the host epithelial cells is likely to involve cooperative binding using both nonspecific and specific interactions including at least two distinct adhesive systems, an RGD-dependent binding of fibronectin mediated by two polypeptide components of the conidial wall of 23 and 30 kD, and a sialic acid-dependent recognition of laminin, fibrinogen and perhaps type IV collagen. Hydrophobic interactions are thought to enable specific short-range interactions to occur by mediating the dehydration and a closer approach of the interacting surfaces. However, the parallel evolution observed between the cell surface hydrophobicity and the expression, or affinity, of the fungal receptors suggests that hydrophobic forces may also play a more direct contributory role in the recognition of the host matrix proteins. Further investigations on the hydrophobic surface components of the conidia as well as cloning and sequencing of the genes encoding the fungal receptors and their comparison with integrins and lectins will provide new insights into a better understanding of the role of these interactions in the adherence of conidia to the host epithelial cells.

References

1 Bodey GP, Vartivarian S: Aspergillosis. Eur J Clin Microbiol Infect Dis 1989;8:413–437.

2 Rinaldi MG: Invasive aspergillosis. Rev Infect Dis 1983;5:1061–1077.

3 Bouchara JP, Tronchin G, Larcher G, Chabasse D: The search for virulence determinants in *Aspergillus fumigatus*. Trends Microbiol 1995;3:327–330.

4 Patti JM, Allen BL, McGavin MJ, Höök M: MSCRAMM-mediated adherence of microorganisms to host tissues. Annu Rev Microbiol 1994;48:585–617.

5 Doolittle RF: Fibrinogen and fibrin. Annu Rev Biochem 1984;53:195–229.

6 Guadiz G, Sporn LA, Simpson-Haidaris PJ: Thrombin cleavage-independent deposition of fibrinogen in extracellular matrices. Blood 1997;90:2644–2653.

7 Beck K, Hunter I, Engel J: Structure and function of laminin: Anatomy of a multidomain glycoprotein. FASEB J 1990;4:148–160.

8 Bouchara JP, Tronchin G, Annaix V, Robert R, Senet JM: Laminin receptors on *Candida albicans* germ tubes. Infect Immun 1990;58:48–54.

9 Bouchara JP, Ait Oumeziane N, Lissitzky JC, Larcher G, Tronchin G, Chabasse D: Attachment of spores of the human pathogenic fungus *Rhizopus oryzae* to extracellular matrix components. Eur J Cell Biol 1996;70:76–83.

10 McMahon JP, Wheat J, Sobel ME, Pasula R, Downing JF, Martin WJ: Murine laminin binds to *Histoplasma capsulatum*: A possible mechanism of dissemination. J Clin Invest 1995;96:1010–1017.

11 Vicentini AP, Gesztesi JL, Franco MF, de Souza W, de Moraes JZ, Travassos LR, Lopes JD: Binding of *Paracoccidioides brasiliensis* to laminin through surface glycoprotein gp43 leads to enhancement of fungal pathogenesis. Infect Immun 1994;62:1465–1469.

12 Pendland JC, Boucias DG: Variations in the ability of galactose and mannose-specific lectins to bind to cell wall surfaces during growth of the insect pathogenic fungus *Paecilomyces farinosus*. Eur J Cell Biol 1993;60:322–330.

13 Pendland JC, Lopez-Lastra C, Boucias DG: Laminin-binding sites on cell walls of the entomopathogen *Nomuraea rileyi* associated with growth and adherence to host tissues. Mycologia 1994; 86:327–335.

14 Vázquez-Juárez R, Ascencio F, Andlid T, Gustafsson L, Wadström T: The expression of potential colonization factors of yeasts isolated from fish during different growth conditions. Can J Microbiol 1993;39:1135–1141.

15 DeHart DJ, Agwu DE, Julian NC, Washburn RG: Binding and germination of *Aspergillus fumigatus* conidia on cultured A549 pneumocytes. J Infect Dis 1997;175:146–150.

16 Weinberger M, Elattar I, Marshall D, Steinberg SM, Redner RL, Young NS, Pizzo PA: Patterns of infection in patients with aplastic anemia and the emergence of *Aspergillus* as a major cause of death. Medicine 1992;7:24–43.

17 Cimon B, Carrère J, Chazalette JP, Giniès JL, Six P, Vinatier JF, Chabasse D, Bouchara JP: Fungal colonization and immune response to fungi in cystic fibrosis. J Mycol Méd 1995;5:211–216.

18 Döring G: The role of proteinases from *Pseudomonas aeruginosa* and polymorphonuclear leukocytes in cystic fibrosis. Drugs Today 1997;33:393–403.

19 Khoo SH, Denning DW: Invasive aspergillosis in patients with AIDS. Clin Infect Dis 1994; 19(suppl 1):S41–S48.

20 Tomee JFC, Wierenga ATJ, Hiemstra PS, Kauffman HF: Proteases from *Aspergillus fumigatus* induce release of proinflammatory cytokines and cell detachment in airway epithelial cell lines. J Infect Dis 1997;176:300–303.

21 Bouchara JP, Bouali A, Tronchin G, Robert R, Chabasse D, Senet JM: Binding of fibrinogen to the pathogenic *Aspergillus* species. J Med Vet Mycol 1988;26:327–334.

22 Annaix V, Bouchara JP, Larcher G, Chabasse D, Tronchin G: Specific binding of human fibrinogen fragment D to *Aspergillus fumigatus* conidia. Infect Immun 1992;60:1747–1755.

23 Tronchin G, Bouchara JP, Larcher G, Lissitzky JC, Chabasse D: Interaction between *Aspergillus fumigatus* and basement membrane laminin: Binding and substrate degradation. Biol Cell 1993;77: 201–208.

24 Tronchin G, Esnault K, Renier G, Filmon R, Chabasse D, Bouchara JP: Expression and identification of a laminin-binding protein in *Aspergillus fumigatus* conidia. Infect Immun 1997;65:9–15.

25 Tronchin G, Bouchara JP, Annaix V, Robert R, Senet JM: Fungal cell adhesion molecules in *Candida albicans*. Eur J Epidemiol 1991;7:23–33.

26 Gill ML, Peñalver MC, Lopez-Ribot JL, O'Connor JE, Martinez JP: Binding of extracellular matrix proteins to *Aspergillus fumigatus* conidia. Infect Immun 1996;64:5239–5247.

27 Tronchin G, Bouchara JP, Latgé JP, Chabasse D: Application of a Lowicryl K4M embedding technique for analysis of fungal adhesins. J Mycol Méd 1993;3:74–78.

28 Coulot P, Bouchara JP, Renier G, Annaix V, Planchenault C, Tronchin G, Chabasse D: Specific interaction of *Aspergillus fumigatus* with fibrinogen and its role in cell adhesion. Infect Immun1994; 62:2169–2177.

29 Bromley IMJ, Donaldson K: Binding of *Aspergillus fumigatus* spores to lung epithelial cells and basement membrane proteins: Relevance to the asthmatic lung. Thorax 1996;51:1203–1209.

30 Peñalver MC, O'Connor JE, Martinez JP, Gil ML: Binding of human fibronectin to *Aspergillus fumigatus* conidia. Infect Immun 1996;64:1146–1153.

31 Tome L, Rex JH, Grazziutti MG, Anaissie EJ, Savary CA: Binding of *Aspergillus fumigatus* (AF) conidia to A549 pneumocytes is inhibited by laminin, fibronectin, and RGDS peptide. Clin Infect Dis 1997;25:378.

32 Lopez-Ribot JL, Casanova M, Monteagudo C, Sepulveda P, Martinez JP: Evidence for the presence of a high-affinity laminin receptor-like molecule on the surface of *Candida albicans* yeast cells. Infect Immun 1994;62:742–746.

33 Ollert MW, Söhnchen R, Korting HC, Ollert U, Bräutigam S, Bräutigam W: Mechanisms of adherence of *Candida albicans* to cultured human epidermal keratinocytes. Infect Immun 1993;61: 4560–4568.

34 Rao CN, Castronovo V, Schmitt MC, Wewer UM, Claysmith AP, Liotta LA, Sobel ME: Evidence for a precursor of the high-affinity metastasis-associated murine laminin receptor. Biochemistry 1989;28:7476–7486.

35 Bouchara JP, Sanchez M, Chevailler A, Marot-Leblond A, Lissitzky JC, Tronchin G, Chabasse D: Sialic acid-dependent recognition of laminin and fibrinogen by *Aspergillus fumigatus* conidia. Infect Immun 1997;65:2717–2724.

36 Ishimaru T, Bernard EM, Tamada S, Armstrong D: The fucose specific lectin (FSL) produced by *Aspergillus fumigatus* may promote attachment of conidia to mammalian cells. Clin Infect Dis 1996;23:898.

37 Silva TMJ, Glee PM, Hazen KC: Influence of cell surface hydrophobicity on attachment of *Candida albicans* to extracellular matrix proteins. J Med Vet Mycol 1995;33:117–122.

38 Thau N, Monod M, Crestani B, Rolland C, Tronchin G, Latgé JP, Paris S: *Rodletless* mutants of *Aspergillus fumigatus*. Infect Immun 1994;62:4380–4388.

39 Parta M, Chang Y, Rulong S, Pinto-DaSilva P, Kwon-Chung KJ: *HYP1*, a hydrophobin gene from *Aspergillus fumigatus* complements the *rodletless* phenotype in *Aspergillus nidulans*. Infect Immun 1994;62:4389–4395.

40 Stringer MA, Dean RA, Sewall TC, Timberlake WE: *Rodletless*, a new *Aspergillus* developmental mutant induced by directed gene inactivation. Genes Dev 1991;5:1161–1171.

41 Stringer MA, Timberlake WE: *dewA* encodes a fungal hydrophobin component of the *Aspergillus* spore wall. Mol Microbiol 1995;16:33–44.
42 Hazen KC: Cell surface hydrophobicity of medically important fungi, especially *Candida* species; in Doyle RJ, Rosenberg M (eds): Microbial Cell Surface Hydrophobicity. Am Soc Microbiol, Washington, 1990, pp 249–295.
43 Ljungh Å, Wadström T: Growth conditions influence expression of cell surface hydrophobicity of staphylococci and other wound infection pathogens. Microbiol Immunol 1995;39:753–757.

J.P. Bouchara, Groupe d'Etude des 'Interactions Hôte-Parasite',
Laboratoire de Parasitologie-Mycologie, Centre Hospitalier Universitaire,
4, rue Larrey, F–49033 Angers Cedex 01 (France)
Tel. +33 2 41 35 34 72, Fax +33 2 41 35 36 16

Brakhage AA, Jahn B, Schmidt A (eds): *Aspergillus fumigatus*.
Contrib Microbiol. Basel, Karger, 1999, vol 2, pp 182–192

Aspergillus fumigatus-Secreted Proteases As Antigenic Molecules and Virulence Factors

Michel Monod [a], *Katia Jaton-Ogay* [b], *Utz Reichard* [c]

[a] Service de Dermatologie (DHURDV), et
[b] Institut de Microbiologie, Centre Hospitalier Universitaire Vaudois, Lausanne, Suisse;
[c] Department of Medical Microbiology, Institute of Hygiene, University of Göttingen, Germany

Aspergillus and *Candida* species are the most frequent pathogenic fungi isolated in hospitals, especially in patients with secondary infections [1]. Aspergilloses are mostly due to *A. fumigatus* and to a lesser extent *A. flavus, A. terreus* and *A. niger. A. fumigatus* is a common soil inhabitant and is found abundantly in decomposing organic matter and composts [2]. In the atmosphere, it releases large amounts of small conidiospores measuring 2–3 μm in diameter [3]. After inhalation of spores, clinical aspergillosis can develop in one of three major manners: (i) allergic bronchopulmonary aspergillosis (ABPA) which occurs when the fungus colonizes the bronchial tree and releases antigenic substances that cause a hypersensitivity pneumonitis; (ii) aspergilloma (fungus ball) that generally develops in pulmonary cavities secondary to other lung diseases such as tuberculosis and anatomic defects, and (iii) invasive pulmonary or disseminated aspergillosis which is the most dangerous form of aspergillosis [1]. Today, even though more patients survive cancer, they are at high risk of fungal infection such as invasive aspergillosis because of the immunosuppressive treatments imposed on these patients. *A. fumigatus* secretes proteases in vitro, and logically, research has been focused on their putative role in virulence similar to the many studies linking secreted proteases and virulence of bacteria.

Endoproteases Secreted by *A. fumigatus*

A. fumigatus secretes three endoproteases: a serine protease referred to as alkaline protease (ALP), a metalloprotease (MEP) and an aspartic protease (PEP) referred to as acid protease. ALP, MEP and PEP are secreted in large amounts when the fungus is cultivated in the presence of protein as the sole nitrogen source [4–12]. The expression of the structural genes of these three proteases is totally inhibited by the presence of free amino acids or small peptides in the growth medium.

ALP was isolated and characterized by several groups [4–8]. This enzyme has a molecular mass of 33 kD, cleaves elastin and collagen and belongs to the subtilisin family. ALP is totally inhibited by phenyl methyl sulfonyl fluoride (PMSF), antipain and chymostatin but not by inhibitors of other serine proteases such as tosyl-*L*-lysine chloromethyl ketone (TLCK) and N-tosyl-*L*-phenylalanine chloromethyl ketone (TPCK). The gene encoding ALP (*ALP*) was isolated [13] and inactivated by targeted gene disruption in pathogenic wild-type strains [14, 15].

MEP was first isolated from ALP nonproducing mutant cultures which still exhibited 30% of the proteolytic activity of wild-type strains [9] and the molecular mass of the enzyme was determined to be 40 kD by SDS-PAGE. MEP cleaves collagen but does not cleave elastin Congo red. However, elastinolytic activity was reported by the detection of radioactivity released after incubation of the enzyme with [^{3}H]elastin [10]. MEP is totally inhibited by specific inhibitors of metalloproteases such as EDTA, 1,10-phenanthroline and phosphoramidon. On the basis of its molecular mass and thermosensitivity, MEP resembles *A. sojae* and *A. oryzae* neutral proteinase I [16, 17]. The gene encoding MEP was cloned and disrupted in wild-type and *ALP* *A. fumigatus* strains [18, 19]. The *alp, mep* double mutant strains were completely devoid of neutral extracellular proteolytic activity in vitro while *mep* *ALP* strains had 70% of the proteolytic activity of the wild-type strain.

PEP has a molecular mass of approximately 38 kD in SDS-PAGE and like all aspartic proteases is totally inhibited by pepstatin [11]. The nucleotide sequence of the gene encoding PEP (*PEP*) [12, 20] shows an exon-intron structure similar to that of the gene encoding the aspartic protease aspergillopepsin O (PEPO) and aspergillopepsin A (PEPA) secreted by *A.* oryzae and *A. niger var awamori*, respectively [21, 22].

ALP, MEP and PEP are not glycosylated. Like many other microbial secreted proteinases, they are synthesized as precursors in the form of the preproproteins. The prepropeptide begins with a signal sequence of 14-21 amino acid containing one to four putative signal peptidase cleavage sites [23]. Their propeptides range from 50 to 226 amino acids long.

Whereas proteolytic activity detected in vitro is due to ALP, MEP and PEP, there is evidence that additional proteases are secreted by *A. fumigatus*:

(i) Immunofluorescence labelling on *A. fumigatus* colonies with anti-PEP polyclonal antibodies showed a surprisingly bright fluorescence on submerged mycelium, on the tips of growing aerial hyphae and on developing conidiophores, without any differences between the wild-type strain and isogenic *pep* mutants [24, 25]. The conidia and the cell wall of the mature aerial hyphae showed no fluorescence. Since there was no difference in this pattern between the wild-type strain and isogenic *pep* mutants the existence of another antigen related to PEP could be assumed. This crossreacting antigen seems to be tightly linked to the cell wall and a common feature of all aspergilli [24]. Correspondingly, the activity of a second aspartyl protease (PEP2), also inhibited by the classical inhibitor pepstatin A, was found in the cell wall fraction of the fungus [25]. However, preliminary results suggest that the latter enzyme is not identical with the antigen responsible for the cell wall fluorescence.

(ii) Genomic DNA encoding a metalloprotease homologous to the neutral protease II of *A. oryzae* has been cloned [26; unpubl. results]. However, no translational product has been detected so far from *A. fumigatus* culture supernatants.

Exoproteases Secreted by *A. fumigatus*

Two dipeptidylpeptidases have been isolated from *A. fumigatus* culture supernatants and characterized at the biochemical and molecular level [27, 28]. The first one releases mainly Ẋ-Ala, but also His-Ser and Ser-Tyr dipeptides from the N-terminal extremity of peptides [27]. This protein was identified in 1967 [29] as one of the two major antigens used in the serodiagnosis of aspergillosis, and was unfortunately called chymotrypsic antigen because of its ability to degrade, like chymotrypsin, the chromogenic substrate N-acetyl-phenylalanine naphtyl ester (NAPNE). Due to its biochemical properties, a new class of dipeptidylpeptidases (DPPV) was created. The second dipeptidylpeptidase releases X-Pro and to a lesser extent X-Ala dipeptides from the N-terminal extremity of peptides and belongs to the class IV of dipeptidylpeptidases (DPPIV) [28]. The DPPV and the DPPIV have an apparent molecular mass of 88 and 95 kD, respectively, and contain approximately 10 kD of N-linked carbohydrates. In contrast to other dipeptidylpeptidases described in humans, mice and yeast, the DPPV and the DPPIV are secreted enzymes with a signal peptide, however, without a prosequence.

Questionable Role of Secreted Proteases in Virulence of *A. fumigatus*

All three *A. fumigatus* endoproteases and both dipeptidylpeptidases are secreted in vivo. Indirect immunofluorescence studies revealed that protease specific antisera label mycelium in the lung of infected mice, as well as of infected patients [4, 10, 11, 30]. In Western blots, ALP did not discriminate between sera samples from aspergilloma patients and other patients not suffering from aspergilloma and allergic bronchopulmonary aspergillosis [30, 31]. The reactivity of control sera against ALP suggested that this protease may share common epitopes with non-*Aspergillus* proteins. In contrast, immunoblot analysis with sera from patients with aspergilloma showed that MEP, PEP and both dipeptidylpeptidases were antigenic and could discriminate between aspergilloma patients and healthy controls [9, 28, 29; unpubl. results].

Elastinolytic activity of *A. fumigatus* has been correlated with the ability of several wild strains to cause invasive aspergillosis in mice [32]. It has also been demonstrated that the proteolytic activity in the brain of infected mice increases proportionally with the growth of *A. fumigatus* [33]. Nevertheless, none of these reports conclusively demonstrated that secreted proteases contribute to the virulence of *Aspergillus*. To determine the importance of the secreted *A. fumigatus* endoproteases in virulence, different mutants were constructed by targeted gene disruption and tested for pathogenicity in murine or guinea pig models [14, 18, 25, 34]. In all experiments with single *alp, mep, pep* mutants and double *alp, mep* mutants (the latter totally deficient in proteolytic activity at neutral pH in vitro) no difference in pathogenicity was observed between the wild-type strains and the mutants. Mortality curves were not statistically different and histopathological studies of lungs from infected mice showed a similar extent of mycelial growth in parental and mutant strains. Invasion of the lung tissues by fungus occurred for the wild-type and mutant strains through the bronchial epithelium after extensive development of the fungus inside the bronchia and bronchioles. When the collagen and elastin matrix was crossed by fungal hyphae no apparent proteolytic degradation was seen around the fungus [18]. Therefore, it must be concluded that the secreted *A. fumigatus* proteolytic activity identified in vitro at neutral and acidic pH, due to *ALP, MEP* and *PEP*, is not essential for the invasion of tissue.

One could object to this somewhat surprising conclusion by a discrepancy between animal models and human infection. The cortisone-pretreated animals were infected only once with a high amount of conidia. In humans, the origin of invasive aspergillosis is the continuous inhalation of conidia released into the air, and the infection only begins when patients are subjected to an immunosuppressive treatment. It is also possible to object that other proteases are produced in vivo by induction through specific host factors. For instance, the three closely related *Candida albicans* aspartyl proteases Sap4, Sap5 and Sap6, which have not yet been detected in cultures in vitro, were produced by *C. albicans* cells following phagocytosis by murine peritoneal macrophages (submitted). Furthermore, the reduction of proteolytic activity could be suppressed by the overexpression of other proteases not expressed, or expressed at a low level, in the wild-type strain. Such mechanisms of compensation have been observed in vitro for the metalloprotease MEP2 in an *alp* mutant of *A. flavus* [35], but not by any proteases in the *A. fumigatus* single *alp, mep* and *pep* mutants we constructed [14, 18, 25]. To definitely abandon the idea that secreted proteolytic activity is related to the pathogenicity of the fungus, it would be necessary to proceed with multiple sequential gene disruptions with the recently developed *PYRG* blaster [36] which can be used like the *URA* blaster in *Saccharomyces cerevisiae* and *C. albicans* [37, 38]. Until now only double mutants have been generated using hygromycin and phleomycin, the only two antibiotics available for such purposes [18].

However, other arguments are in favor for the absence of proteases in pathogenicity of *A. fumigatus*: (i) the lack of vessel wall elastolysis in human invasive aspergillosis demonstrated by Denning et al. [39] with tissues obtained at autopsy; (ii) the fungal virulence not attenuated by monoclonal antibodies able to inhibit ALP [40]; (iii) the production of homologous enzymes to ALP and PEP by many fungi which are not pathogenic. Consequently, if it is reasonable to admit that secreted proteolytic activity is not sufficient to explain the pathologic behavior of *A. fumigatus*, it is necessary to search elsewhere for other factors to elucidate the development of the fungus in the human host. *A. fumigatus* is the archetype of opportunistic fungus which is saprophytic in the environment, and does not invade immunocompetent people. This fungus grows rapidly at 37 °C on rich and minimal medium. One could consider that *A. fumigatus* is capable of finding a particular ecological niche in the bronchia. This niche would be characterized by nutrient and temperature conditions favorable to its saprophytic growth since this fungus accounts for only a small proportion of the airborne mold spores, even though it is the most frequently isolated fungus from lung and sputum [41, 42]. Infection of lung tissues by *A. fumigatus* occurs when the host phagocytic defences have been weakened by immunosuppressive treatments.

Expression of *A. fumigatus*-Secreted Proteases in *Pichia pastoris*

Numerous antigens or semipurified antigenic fractions from *A. fumigatus* have been described in the literature but only a few have been purified and

Table 1. Construction of plasmids for the expression of *ALP, MEP* and *PEP* in *P. pastoris*

Genes	DNA sequences and accession numbers [ref.]		Oligonucleotide primers	Amino acid sequence	PCR products (with cloning sites) inserted in pKJ111 and pKJ1133	*P. pastoris-E. coli* shuttle vector used and cloning sites
MEP	Z30424 [18]	sense	CTG *TTT/AAA* CAT CCC GCT CAC CAG TCT TAC	(K)H P A H Q S Y	*MEP*(bp55-1899)	pKJ111
		antisense	GTC *TTT/AAA* **CTA** ACA GAC ACC ACT GGG GAC CTC	E V P S G V C **STOP**	(*Dra*I-*Dra*I)	*Sma*I
ALP	Z11580 [13]	sense	CT G*C/T CGA G*CG CCT GTC CAG GAA ACT CGT	(R)A P V Q E T R	ALP(bp61-1203)	pKJ113
		antisense	AGC *G/GA TCC* **CTA** AGC ATT GCC ATT GTA GGC AAG	L A Y N G N A **STOP**	(*Xho*I-*Bam*HI)	(*Xho*I-*Bam*HI)
PEP	X85092 [20]	sense	A T*C/T CGA G*CT GTC CCT GTG GTC CAG CCG CGC	(R)A V P V V Q P R	*PEP*(bp61-1185)	pKJ113
		antisense	T A*G/G ATC* **CTA** TGC CTG AGG GGC GAA GCC	G F A P Q A **STOP**	(*Xho*I-*Bam*HI)	(*Xho*I-*Bam*HI)

Expression plasmids were constructed by cloning a PCR product of *A. fumigatus* genes in the multiple cloning site of the *E. coli-P. pastoris* shuttle vector pKJ111 or pKJ113 (fig. 1). PCR was performed using homologous primers derived from *MEP, ALP* and *PEP*, and *A. fumigatus* cDNA as template. After purification, the PCR products were digested by restriction enzymes for which a site was previously designed at the 5′ extremity of the primers and cloned into the appropriate sites of the multiple cloning site of the *E. coli-P. pastoris* shuttle vector. *P. pastoris* GS115 was transformed by preparing spheroplasts as described in the Manual Version 2.0 of the *Pichia* Expression Kit (Invitrogen) with 10 μg of linearized plasmid DNA. Transformants selected on histidine-deficient medium were screened for insertion of the construct at the *AOX1* site on minimal methanol plates. Transformants unable to grow on media containing only methanol as a carbon source were assumed to contain the construct at the correct yeast genomic location by integration events in the *AOX1* locus displacing the *AOX1* coding region. These transformants were further screened for enzyme production in methanol-containing medium.

characterized biochemically and immunologically. Secreted proteases have proven to be antigenic and useful in order to detect aspergilloma and ABPA [9; unpubl. results]. In this respect, *Pichia pastoris* was chosen as an expression system to produce substantial amounts of highly purified antigens.

Whereas the signal peptide is necessary for entering the secretory pathway by transporting the protein across the membrane of the endoplasmic reticulum, the propeptide of many microbial proteinases have been found to be essential and specific for assisting the correct folding and the secretion of their associated

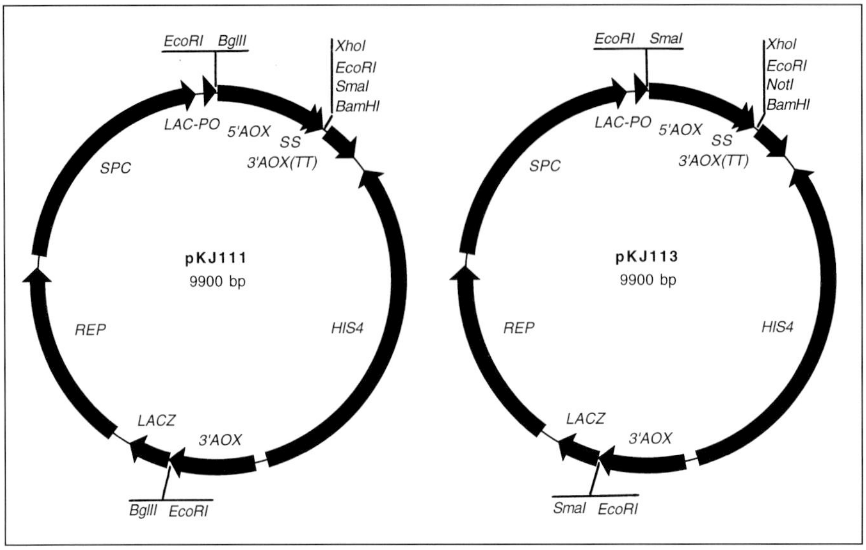

Fig. 1. Map of plasmid pKJ111 and pKJ113. *REP:* pCL1920 replicon [44]; *SPC:* Spectinomycin resistance gene; *LACPO: LACZ* promoter; *5'AOX: P. pastoris* alcohol oxidase gene (*AOX*) promoter; *SS:* signal peptide sequence of the *P. pastoris* acid phosphatase gene (*PHO1*); *3'AOX(TT): AOX* terminator; *HIS4: P. pastoris* histidinol dehydrogenase gene; *3'AOX: 3'AOX* downstream sequence; *LACZ:* pUC19 *LACZ* fragment.

proteins [43]. Initially, we attempted to clone a cDNA fragment MEP encoding *A. fumigatus* MEP with its propeptide (bp 55-1899; table 1) into the pHILS1 expression vector (Invitrogen) downstream of the *P. pastoris* acid phosphatase gene (*PHO1*) signal sequence. Unfortunately, no correct construction was obtained. Preliminary experiments have shown that fungal proteinase DNA can be successfully cloned in *Escherichia coli* only by using a low copy number vector. We therefore cloned the expression cassette of pHILS1 into derivatives of the low copy number plasmid pCL1920 [44], and constructed two new vectors, pKJ111 and pKJ113 (fig. 1). In pKJ111, the pHILS1 expression cassette was flanked by two *Bgl*II sites for linearization of the DNA before transformation of *P. pastoris.* In pKJ113, the pHILS1 expression cassette, in which a *Not*I site replaces the *Sma*I site in the multiple cloning site, was flanked by two *Sma*I sites. The cloning of DNA fragments encoding MEP with its prosequence was successfully accomplished in pKJ111. Subsequently, the *his MUT P. pastoris* strain GS115 was used as a host for transformation with linearized plasmid DNA, and *HIS mut* transformants were selected and screened for protease production. *A. fumigatus* MEP could be recovered in correctly processed form from culture supernatant of *P. pastoris* transformed

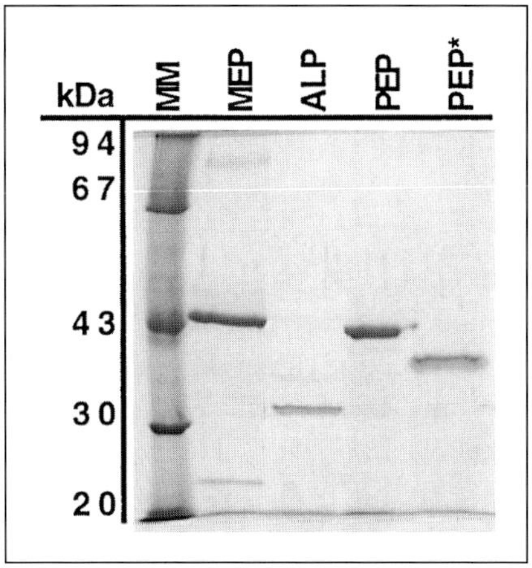

Fig. 2. Protein profile of the culture supernatant of *P. pastoris* strains producing ALP, MEP and PEP. PEP* is for recombinant PEP in *P. pastoris* supernatant buffered at pH 3.0. The proteins in 10 µl of supernatant were loaded without further treatment onto 12% SDS-PAGE. The gel was stained with Coomassie brillant blue. Molecular masses of protein standards are shown in the left margin.

with pSB1, with a yield of 40 µg/ml and was active. Southern blot analysis of the DNA of the transformants secreting MEP showed that the pHILS1 expression cassette was inserted in the *P. pastoris* genome at the AOX1 locus via homologous recombination (data not shown).

In order to express *A. fumigatus* ALP and PEP, cDNA fragments encoding the proteins with their prosequence were cloned in pKJ113 (table 1). The activity and molecular mass of ALP secreted by *P. pastoris HIS mut* transformants and ALP secreted by *A. fumigatus* were similar. In contrast a protein 4 kD longer than mature PEP from *A. fumigatus* was found in *P. pastoris* culture supernatant suggesting that an unprocessed form of the protein was secreted at neutral pH. Active and correctly processed PEP could be recovered from *P. pastoris* culture supernatant when the latter was buffered at pH 3.0 (fig. 2). Expression of active recombinant DPPV and DPPIV was also successful in *P. pastoris* with a yield of 150 µg/ml [26, 27].

The availability of *A. fumigatus*-secreted proteases from *P. pastoris* will assist future planned studies in the fields of protease biochemistry, cristallography and clinical microbiology. Expression of active enzymes in a heterologous system bypasses the problem of cross-reaction with impurities from *A. fumigatus* not removed after laborious purification of the proteins. Furthermore, expression in *P. pastoris* provides antigen in substantial amounts. For instance, where 0.5 mg of MEP would be extracted from one liter of culture supernatant of *A. fumigatus* [12], the same amount could be extracted from 10 ml of culture supernatant of *P. pastoris*. The use of these antigens prepared from *P. pastoris* is now being tested within a large population of patients suffering from different forms of aspergillosis.

Conclusions

A. fumigatus, the species most frequently isolated from patients with aspergillosis, secretes serine, metallo- and aspartic proteases. Recently, two dipeptidylpeptidases have been isolated from *A. fumigatus* culture supernatants and characterized. The role of the secreted proteases in the virulence of *A. fumigatus* is questionable since no significant difference for pathogenicity was observed with protease mutants obtained by targeted mutagenesis and the parental strain. Recombinant proteases and dipeptidylpeptidases expressed in *P. pastoris* are enzymatically active and have biochemical and antigenic properties similar to the protein isolated from *A. fumigatus*. The availability of these recombinant proteins will assist future studies in clinical microbiology to improve the diagnosis of allergic aspergillosis and aspergilloma.

References

1 Bodey GP, Vartivarian S: Aspergillosis. Eur J Clin Microbiol Infect Dis 1989;8:413–437.

2 Haines J: *Aspergillus* in compost: Straw man or fatal flaw. Biocycle1995;32–35.

3 Samson RA, Van Reenen-Hoekstra ES: Introduction to Food-Borne Fungi. Baarn, Delft, Centraalbureau voor Schimmelcultures, 1988.

4 Reichard U, Büttner S, Eiffert H, Staib F, Rüchel R: Purification and characterisation of an extracellular serine proteinase from *Aspergillus fumigatus* and its detection in tissue. J Med Microbiol 1990;33:243–251.

5 Monod M, Togni G, Rahalison L, Frenk E: Isolation and characterisation of an extracellular alkaline protease of *Aspergillus fumigatus*. J Med Microbiol 1991;35:23–28.

6 Kolattukudy PE, Lee JD, Rogers LM, Zimmerman P, Ceselski S, Fox B, Stein B, Copelan EA: Evidence for possible involvement of an elastolytic serine protease in aspergillosis. Infect Immun 1993;61:2357–2368.

7 Frosco MB, Chase T, MacMillan JD: Purification and properties of the elastase from *Aspergillus fumigatus*. Infect Immun 1992;60:728–734.

8 Larcher G, Bouchara JP, Annaix V, Symoens F, Chabasse D, Tronchin G: Purification and characterization of a fibrinogenolytic serine proteinase from *Aspergillus fumigatus* culture filtrate. FEBS Lett 1992;308:65–69.

9 Monod M, Paris S, Sanglard D, Jaton-Ogay K, Bille J, Latgé JP: Isolation and characterization of a secreted metalloprotease of *Aspergillus fumigatus*. Infect Immun 1993;61:4099–4104.

10 Markaryan A, Morozova I, Yu H, Kolattukudy PE: Purification and characterization of an elastinolytic metalloprotease from *Aspergillus fumigatus* and immunoelectron microscopic evidence of secretion of this enzyme by the fungus invading the murine lung. Infect Immun 1994;62:2149–2157.

11 Reichard U, Eiffert H, Rüchel R: Purification and characterization of an extracellular aspartic proteinase from *Aspergillus fumigatus*. J Med Vet Mycol 1994;32:427–436.

12 Lee JD, Kolattukudy PE: Molecular cloning of the cDNA and gene for an elastinolytic aspartic proteinase from *Aspergillus fumigatus* and evidence of its secretion by the fungus during invasion of the host lung. Infect Immun 1995;63:3796–3803.

13 Jaton-Ogay K, Suter M, Crameri R, Falchetto R, Fatih A, Monod M: Nucleotide sequence of a genomic and a cDNA clone encoding an extracellular alkaline protease of *Aspergillus fumigatus*. FEMS Microbiol Lett 1992;92:163–168.

14 Monod M, Paris S, Sarfati J, Jaton-Ogay K, Ave P, Latgé JP: Virulence of alkaline protease-deficient mutants of *Aspergillus fumigatus*. FEMS Microbiol Lett 1993;106:39–46.

15 Tang CM, Cohen J, Holden W: An *Aspergillus fumigatus* alkaline protease mutant constructed by gene disruption is deficient in extracellular elastase activity. Mol Microbiol 1993;6:1663–1671.
16 Sekine H: Neutral proteinases I and II of *Aspergillus sojae*. Isolation in homogeneous form. Agric Biol Chem 1972;36:198–206.
17 Nakadai T, Nasuno S, Iguchi N: Purification and properties of neutral proteinase I from *Aspergillus oryzae*. Agric Biol Chem 1973;37:2695–2701.
18 Jaton-Ogay K, Paris S, Huerre M, Quadroni M, Falchetto R, Togni G, Latgé JP, Monod M: Cloning and disruption of the gene encoding an extracellular metalloprotease of *Aspergillus fumigatus*. Mol Microbiol 1994;14:917–928.
19 Sirakova TD, Markaryan A, Kolattukudy PE: Molecular cloning and sequencing of the cDNA and gene for a novel elastinolytic metalloproteinase from *Aspergillus fumigatus* and its expression in *Escherichia coli*. Infect Immun 1994;62:4208–4218.
20 Reichard U, Monod M, Rüchel R: Molecular cloning and sequencing of the gene encoding an extracellular aspartic proteinase from *Aspergillus fumigatus*. FEMS Microbiol Lett 1995;130:69–74.
21 Berka RM, Carmona CL, Hayenga KJ, Thompson SA, Ward M: Isolation and characterization of the *Aspergillus oryzae* gene encoding aspergillopepsin O. Gene 1993;125:195–198.
22 Berka RM, Ward M, Wilson LJ, Hayenha KJ, Kodama KH, Carlomango LP, Thompson SA: Molecular cloning and deletion of the gene encoding aspergillopepsin A from *Aspergillus awamori*. Gene 1990;86:153–162.
23 Von Heijne G: A new method for predicting signal sequence cleavage sites. Nucl Acids Res 1986; 14:4683–4690.
24 Reichard U, Monod M, Rüchel R: Expression pattern of aspartic proteinase antigens in aspergilli. Mycoses 1996;39:99–101.
25 Reichard U, Monod M, Odds F, Rüchel R: Virulence of an aspergillopepsin-deficient mutant of *Aspergillus fumigatus* and evidence for another aspartic proteinase linked to the fungal cell wall. J Med Vet Mycol 1997;35:189–195.
26 Ramesh MV, Sirakova TD, Kolattukudy PE: Cloning and characterization of the cDNAs and genes (*mep20*) encoding homologous metalloproteinases from *Aspergillus flavus* and *A. fumigatus*. Gene 1995;165:121–125.
27 Beauvais A, Monod M, Debeaupuis JP, Diaquin M, Kobayashi H, Latgé JP: Biochemical and antigenic characterization of a new dipeptidyl-peptidase isolated from *Aspergillus fumigatus*. J Biol Chem 1997;272:6238–6244.
28 Beauvais A, Monod M, Wyniger J, Debeaupuis JP, Grouzmann E, Brakch N, Svab J, Hovanessian AG, Latgé JP: Dipeptidyl-peptidase IV secreted by *Aspergillus fumigatus*, a fungus pathogenic to humans. Infect Immun 1997;67:3042–3047.
29 Biguet J, Tran Van Ky J, Andrieu S: Identification d'une activité chymotrypsique au niveau de fractions remarquables d'*Aspergillus fumigatus*. Répercussions sur le diagnostic immunologique de l'aspergillose. Rev Immunol Paris 1967;31:317–328.
30 Moutaouakil M, Monod M, Prevost MC, Bouchara JP, Latgé JP: Identification of the 33-kDa alkaline protease of *Aspergillus fumigatus* in vitro and in vivo. J Med Microbiol 1993;39:393–399.
31 Moser M, Menz G, Blaser K, Crameri R: Recombinant expression and antigenic properties of a 32-Kilodalton extracellular alkaline protease, representing a possible virulence factor from *Aspergillus fumigatus*. Infect Immun 1994;62:936–942.
32 Kothary MH, Chase T, MacMillan JD: Correlation of elastase production by some strains of *Aspergillus fumigatus* with ability to cause pulmonary invasive aspergillosis in mice. Infect Immun 1984;43:320–325.
33 Miyaji M, Nishimura K: Relationship between proteolytic activity of *Aspergillus fumigatus* and the fungus invasiveness of mouse brain. Mycopathologia 1977;62:161–166.
34 Tang CM, Cohen J, Krausz T, Van Noorden S, Holden DW: The alkaline protease of *Aspergillus fumigatus* is not a virulence determinant in two murine models of invasive pulmonary aspergillosis. Infect Immun 1993;61:1650–1656.
35 Ramesh MV, Kolattukudy PE: Disruption of the serine proteinase gene (*sep*) in *Aspergillus flavus* leads to a compensatory increase in the expression of a metalloproteinase gene (*mep20*). J Bacteriol 1996;178:3899–3907.

36 D'Enfert C: Selection of multiple disruption events in *Aspergillus fumigatus* using the orotidine-5′-decarboxylase gene, *pyrG*, as a unique transformation marker. Curr Genet 1996;30:76–82.
37 Fonzi WA, Irwin MY: Isogenic strain construction and gene mapping in *Candida albicans*. Genetics 1993;134:717–728.
38 Sanglard D, Hube B, Monod M, Odds FC, Gow NAR: A triple deletion of the secreted aspartyl proteinase genes *SAP4, SAP5,* and *SAP6* of *Candida albicans* causes attenuated virulence. Infect Immun 1997;65:3539–3546.
39 Denning DW, Ward PN, Fenelon LE, Benbow EW: Lack of vessel wall elastolysis in human invasive aspergillosis. Infect Immun 1992;60:5153–5156.
40 Frosco MT, Chase T, MacMillan JD: The effect of elastase-specific monoclonal and polyclonal antibodies on the virulence of *Aspergillus fumigatus* in immunocompromised mice. Mycopathologia 1994;125:65–76.
41 Mullins J, Seaton A: Fungal spores in lung and sputum. Clin Allergy 1978;8:525–533.
42 Schmitt JJ, Blevins A, Sobeck K, Amstrong D: *Aspergillus* species from hospital air and from patients. Mycoses 1991;33:539–541.
43 Eder J, Fersht AR: Pro-sequence-assisted protein folding. Mol Microbiol 1995;16:609–614.
44 Lerner CG, Inouye M: Low copy number plasmids for regulated low-level expression of cloned genes in *Escherichia coli* with blue/white insert screening capability. Nucl Acids Res 1990;18:4631.

Dr. Michel Monod, Laboratoire de Mycologie, Service de Dermatologie,
Hôpital de Beaumont 04–423, Centre Hospitalier Universitaire Vaudois,
CH–1011 Lausanne (Switzerland)
Tel. +41 21 314 0376, Fax +41 21 314 0378, E-Mail Michel.Monod@chuv.hospvd.ch

Brakhage AA, Jahn B, Schmidt A (eds): *Aspergillus fumigatus*.
Contrib Microbiol. Basel, Karger, 1999, vol 2, pp 193–204

Chitin Synthase Genes of *Aspergillus* Species

Hiroyuki Horiuchi, Masamichi Takagi

Department of Biotechnology, The University of Tokyo, Japan

Introduction

Chitin is a polymer of β1,4-linked *N*-acetyl glucosamine and the major cell wall component of many filamentous fungi. Since chitin is believed to have a very rigid structure, it is generally accepted that chitin represents an important determinant for forming the shapes of hyphae and conidiophores. Furthermore, it should play an important role in tip growth, branching of hyphae, and differentiation of conidiophores. Therefore, chitin synthases of filamentous fungi would be interesting targets for antifungal drugs. Thus, we have been focusing on the synthesis as well as the degradation of chitin in filamentous fungi, especially *Aspergillus nidulans* as a model organism for pathogenic species such as *Aspergillus fumigatus*.

Chitin synthases (EC 2.4.1.16) catalyse the polymerization of *N*-acetyl-*D*-glucosamine (GlcNAc), and have been split into five classes (I–V) according to their structural characteristics (fig. 1). Each of the chitin synthases is composed of three regions: a hydrophilic nonconserved region at the N-terminus, a neutral highly conserved region, and a hydrophobic region at the C-terminus, which is thought to anchor the enzyme of the membrane. The N-terminal and the central regions are suggested to face the cytoplasmic side.

The study of chitin synthase genes and their gene products is most advanced for *Saccharomyces cerevisiae*. Until now, three chitin synthase structural genes have been cloned from this yeast and designated *CHS1, CHS2* and *CHS3* [1]. The products of both *CHS2* (Chs2p) and *CHS3* (Chs3p) belong to classes II and IV, respectively. The product of *CHS1* (Chs1p) however, does not belong to any of the five classes (fig. 1). The functions of these gene products have been investigated in detail. Chs1p functions as a repair enzyme

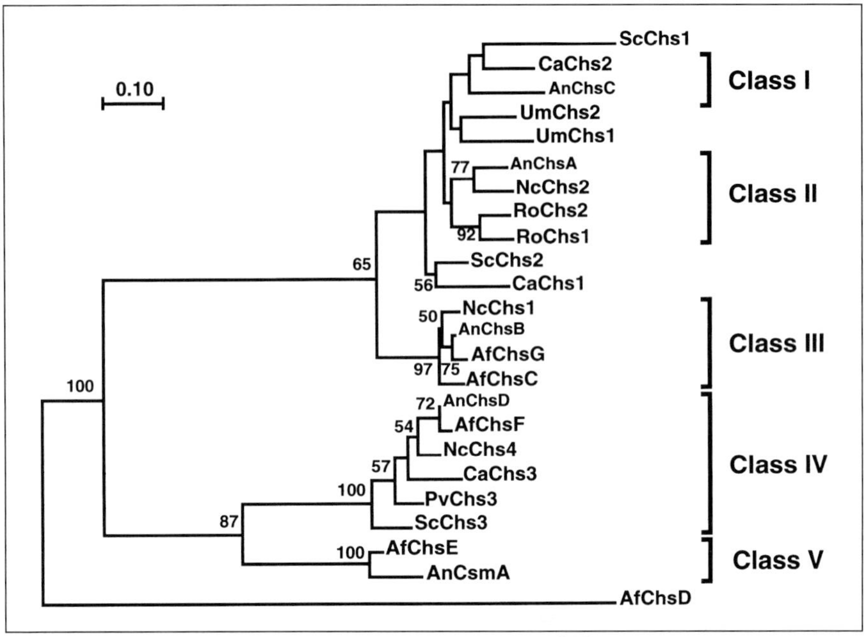

Fig. 1. A phylogenetic tree based on conserved amino acid sequences in chitin synthases. The scale bar denotes 0.10 substitution per site. The tree was constructed by the neighbor-joining method, and bootstrap values above 50 from 100 resamplings are shown for each node. Sc = *Saccharomyces cerevisiae*; Ca = *Candida albicans*; An = *A. nidulans*; Um = *Ustilago maydis*; Nc = *Neurospora crassa*; Ro = *Rhizopus oligosporus*; Af = *A. fumigatus*; Pv = *Phialophore verrucosa*.

when chitinase (Cts1p) has digested chitin too extensively at the mother-daughter neck region [2]. The function of Chs2p is to synthesize the primary septum between mother and daughter cells. Analysis of the structure-function relationship of Chs2p suggested that its active center is located in the central conserved region [3]. The functions of the N-terminal region of both Chs1p and Chs2p are unclear [4, 5]. Chs3p synthesizes most of the cell wall chitin (~90%) [1]. For the proper localization of Chs3p, the gene products of both *CHS4* [6] and *CHS5* [7] are required. Synthesis of Chs1p and Chs3p is constitutive, whereas that of Chs2p is cell-cycle regulated. These three chitin synthases are transported to the plasma membrane via the secretory pathway and then internalized via the endocytotic pathway. Chs2p is then rapidly degraded, while Chs1p and Chs3p may be recycled to the plasma membrane [8, 9]. These results suggest that each chitin synthase has a different function at a different stage of the cell cycle or at a different site of the cell.

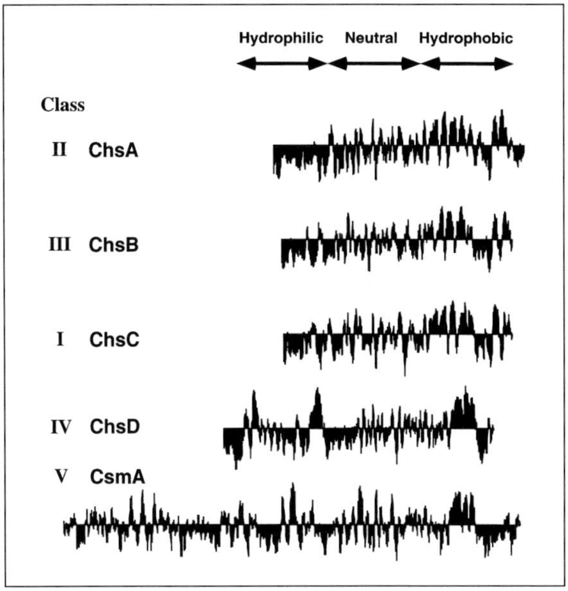

Fig. 2. Hydropathy profiles of chitin synthases. The N-terminal hydrophilic region, the central neutral region, and the C-terminal hydrophobic region are indicated by two-headed arrows.

So far we have cloned from *A. nidulans* four chitin synthase genes (*chsA, chsB, chsC* and *chsD*) and one gene (*csmA*) which encodes chitin synthase with a myosin motor-like domain fused at the N-terminus. In this section, we mainly describe the structures and functions of the gene products of these five genes.

Cloning and Structures of Chitin Synthase Genes from *A. nidulans* and *A. fumigatus*

chsA and *chsB* were cloned from the genomic library of *A. nidulans* by low-stringent hybridization with *CHS2* as a probe [10]. Partial fragments of *chsC, chsD* and *csmA* were amplified by polymerase chain reaction using degenerate oligonucleotides as primers [11–13]. These three genes were then cloned in full length from the genomic library. The DNA sequences of all five genes were determined and the positions of the introns were determined either by the RACE method or predicted from the positions of intron-specific consensus splice sites characteristic of filamentous fungi. Surprisingly, *csmA*

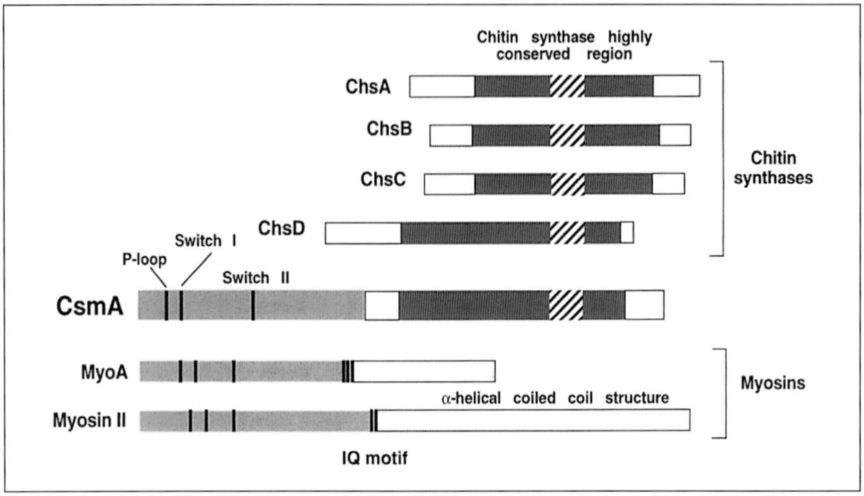

Fig. 3. Structures of *CsmA*, chitin synthases of *A. nidulans*, and two myosins. Comparison of the structure of CsmA with other related proteins. ChsA, ChsB, ChsC and ChsD, from *A. nidulans*; MyoA, from *A. nidulans*. Myosin II = Chicken muscle myosin II.

seems to encode a chitin synthase with a myosin motor-like domain fused at its N-terminus, thus, instead of *chsE*, we named it chitin synthase with a myosin motor-like domain. Deduced products of these genes were designated ChsA, ChsB, ChsC, ChsD, and CsmA, respectively. *chsD* and a partial sequence of *csmA* were also isolated by Specht et al. [14]. They designated our *chsD* as *chsE* and the 3′ half of *csmA* as *chsD*. Here, we refer to their *chsE* and *chsD* as *'chsE'* and *'chsD'*, respectively. Each chitin synthase (ChsA ~ ChsD) and chitin synthase domain of CsmA can be classified into a different group. ChsA, ChsB, ChsC, ChsD, and the chitin synthase domain of CsmA belong to classes II, III, I, IV, and V, respectively. The phylogenetic tree drawn from their amino acid sequences is shown in figure 1. Their hydropathy profiles indicate that each has a typical profile specific of members of each class (fig. 2).

The structure of CsmA is aligned with the four other synthases of *A. nidulans* isolated in our laboratory and with myosins from various organisms reported in the literature (fig. 3). In addition to overall similarity between the myosin motor-like domain of CsmA and other myosins, the positions of the P-loop, the switch I, the switch II, and the essential residues in these loops are conserved among them.

Mellado et al. [15] have succeeded in isolating seven fragments encoding partial sequences of chitin synthases by PCR amplification using the genomic

Table 1. Comparison of amino acid sequences of chitin synthases and chitin synthase domains

	ChsA	ChsB	ChsC	ChsD	CsmAcs*	AfChsC	AfChsD	AfChsE
ChsB	37.0							
ChsC	41.0	39.8						
ChsD	19.8	20.4	20.2					
CsmAcs	21.4	19.5	20.8	30.5				
AfChsC	37.0	67.9	39.7	20.7	20.5			
AfChsD	21.0	20.8	19.7	20.6	21.5	20.9		
AfChsE	21.8	21.3	21.9	27.6	48.3 (76.5)	21.3	19.4	
AfChsG	37.7	88.9	39.5	19.4	18.6	68.3	20.2	20.7

* Chitin synthase domain of CsmA (C-terminal 952 amino acids of CsmA).
Similarity between C-terminal 1,543 amino acids of CsmA and AfChsE is shown in parenthesis. Af = *A. fumigatus.*

DNA of *A. fumigatus* as a template and named them *chsA*∼*chsG* (we refer to them as Af*chsA*∼Af*chsG* here). So far, four of these sequences have been determined in full length and reported [16–18]. The similarity of deduced amino acid sequences of chitin synthases (or chitin synthase domains) between *A. nidulans* and *A. fumigatus* is summarized in table 1. Interestingly, the structures of the gene products of Af*chsE* (AfChsE) and Af*chsG* (AfChsG) are highly similar to the chitin synthase domains of CsmA and ChsB, respectively. Characterization of chitin synthase activity of ChsB was done by Tatsuno et al. [19].

Expression of *chs* Genes Analyzed by Using the *Escherichia coli* β-Galactosidase Gene as a Reporter

To determine the expression pattern of *chs* genes, the N-terminal part of *chsA, chsB, chsC* and *chsD* was fused with the β-galactosidase gene (*lacZ*) of *Escherichia coli* as a translational fusion and transformed into *A. nidulans.* Transformants in which only one copy of a gene fusion was integrated into the *chsA* locus (in the case of *chsA-lacZ*) or the *argB* locus were selected and designated AG, BG, CG and DG strains for *chsA, chsB, chsC* and *chsD*, respectively. In situ staining experiments of these transformants showed that *chsA* had very low expression in hyphae. Its expression was increased in

4a 4b 4c

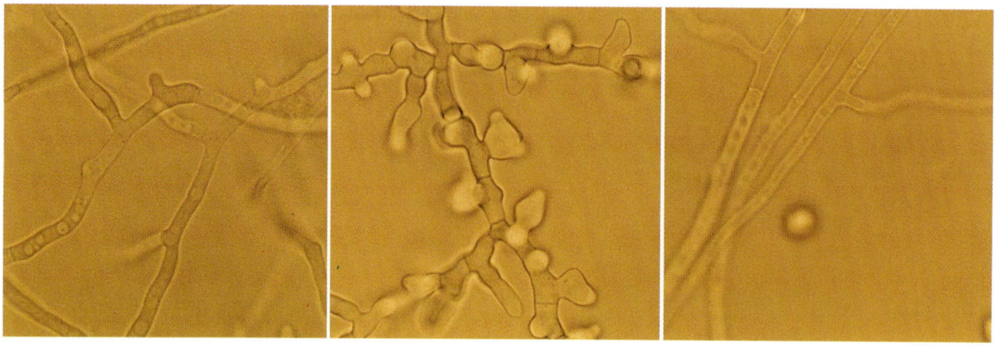

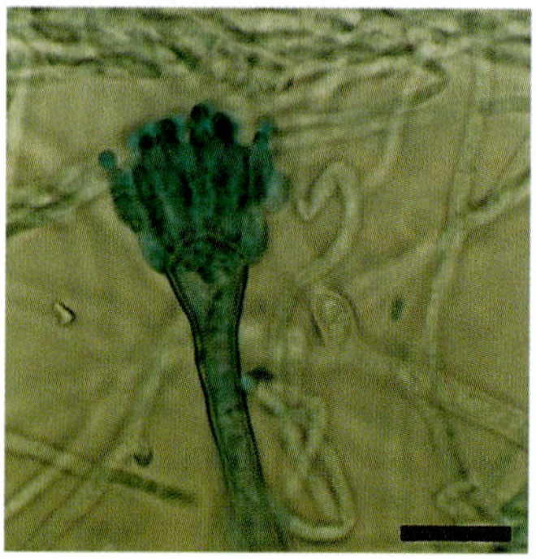

5

Fig. 4. Morphology of hyphae of the *chsB*-conditional mutant. The wild-type strain, the *chsB*-conditional mutant, and the *chsB*-conditional mutant containing an extra copy of *chsB* were grown on MMG medium for 24 hr at 37 °C. *a* The wild-type strain. *b* The *chsB*-conditional mutant. *c* The *chsB*-conditional mutant containing an extra copy of *chsB*.

Fig. 5. In situ staining of *A. nidulans* transformed with the *chsC-lacZ* gene fusion. The bar represents 10 μm.

metulae, phialides and conidia. Expression of *chsB* was also increased in metulae, phialides and conidia but in contrast to the case of *chsA*, moderate staining was also detected in hyphae. Induction of expression of *chsA* and *chsB* during conidia formation was confirmed by measuring β-galactosidase activity in AG and BG strains. The whole activity of β-galactosidase in the BG strain was higher than that in the AG strain throughout the culture period.

The expression patterns of both *chsC* (fig. 4) and *chsD* were similar to *chsB*. The characteristics of structures and expression patterns of *chsA-D* and *csmA* are summarized in table 2.

Table 2. Properties of chitin synthases and chitin synthase-like protein of *A. nidulans*

	ChsA	ChsB	ChsC	ChsD	CsmA
Class	II	III	I	IV	V
No. of a.a. residues	1,013	916	914	1,086	1,852
Expression	conidiophores (hyphae)	conidiophores hyphae	conidiophores hyphae	conidiophores hyphae	n.d.
Deduced functions	conidiophore formation	hyphal growth	conidiophore formation	conidiophore formation	hyphal growth
	(hyphal growth)	(conidiophore formation)	(hyphal growth)	(hyphal growth)	conidiophore formation

n.d. = Not determined.

Single Disruption of *chs* Genes

For the understanding of the function of a gene and its product, the gene disruption technique has widely been used. Therefore, we constructed strains of *A. nidulans* in which one of four *chs* genes or the *csmA* gene was disrupted using the *argB* gene of *A. nidulans* or the *pyr4* gene of *Neurospora crassa* as selectable markers. Compared with the parental strain ABPU1 (*biA1 pyrG89; wA3; argB2; pyroA4*), no phenotypic change was detected in *chsA, chsC*, or *chsD* disruptant strains under various conditions [10–12]. On the contrary, the *chsB* (class III) disruptant strain exhibited a severe defect in hyphal growth, which was not mediated again by the addition of 1.2 *M* sorbitol into the medium as an osmotic stabilizer [10]. This finding suggested that ChsB has an essential function in normal tip growth. In *N. crassa*, mutants of both *chs-2* and *chs-4* which encode class II and IV chitin synthases, respectively, grew and developed normally under standard conditions [20, 21], while the mutation of *chs-1* encoding a class III chitin synthase, caused severe defects in hyphal growth [20]. Two genes (Af*chsC* and Af*chsG*) encoding class III chitin synthases were cloned from *A. fumigatus* and disrupted. Whereas the Af*chsC* disruptant showed no phenotypic change, the Af*chsG* disruptant strain showed a reduced growth rate and formed highly branched hyphae. These phenotypes were not suppressed by the addion of 1 *M* sorbitol to the media [16]. Taken together, these results suggest that most class III chitin synthases have important roles in tip growth of hyphae in filamentous fungi. Since no class III chitin synthase has been isolated from yeasts so far, it is possible that this type of chitin synthase is specific and very important for the hyphal growth of filamentous fungi. The presence of chitin synthases of this class in filamentous fungi only may be one of the causes of different cell shapes between filamentous fungi and yeasts.

The open reading frame of *chsB* was placed under the control of the *alcA* promoter [*alcA* encodes alcohol dehydrogenase]. Its strong promoter can be regulated by changing carbon sources, e.g. glucose represses and threonine induces the expression [23]. This gene fusion was introduced into *A. nidulans*. Transformants in which the *chsB* gene on the chromosome was replaced by the gene fusion were selected (*chsB*-conditional mutants). These could grow as fast as the wild-type strain when the *alcA* promoter was turned on on minimal medium containing threonine and fructose as the carbon sources (MMFT medium). However, they grew very weakly when the promoter was turned off on minimal medium containing glucose as the sole carbon source (MMG medium) (fig. 4). Under the latter condition, hyphae of these strains were hyperbranched and had abnormal structures. These characteristics were remedied when another copy of *chsB* with its own promoter was introduced into these strains (fig. 4, 5).

To determine the conidiation efficiency of the *chsB*-conditional mutant, we carried out the following experiment. The mutant and the wild-type strains were grown in liquid MMFT medium for 17 h to allow expression of *chsB* and normal tip growth, and then the strains were transferred onto the solid MMG medium for 25 h to induce conidia formation in the absence of expression of *chsB*, and then the conidiation efficiency was measured. The result showed that the conidiation efficiency of the mutant was approximately 45% of that of the wild-type strain [Motoyama et al., unpubl. results]. Assuming that the ChsB synthesized during liquid culture remains and functions in the cells during conidia formation, a 55% reduction in the conidiation efficiency would be the minimum estimation of the effect of *chsB* repression on conidiation. Thus, it seems likely that ChsB also has some functions in conidia formation. Borgia et al. reported that the *chsB* disruptants constructed by them did not form aerial hyphae, conidiophores, or conidia [24].

The *csmA* (class V) disruptants were constructed from ABPU1 using *argB* as the selectable marker. These disruptant strains exhibited some phenotypic changes. The colony growth rate of the disruptants was almost the same as that of the wild-type strain, but was severely inhibited in the presence of chitin-binding reagents, such as Calcofluor white and Congo red. Abnormal morphology of hyphae, septa and conidiophores was observed by microscopy. Moreover, so-called intrahyphal hyphae, i.e. hyphae produced inside hyphae, were formed occasionally. These results indicate that *csmA* has important roles in the polarized synthesis and maintenance of integrity of the cell wall [Fujiwara et al., in prep.]. Specht et al. [14] reported that disruption of '*chsD*' caused a severe growth defect of disruptants when grown on the low osmotic medium. This effect was suppressible by the addition of KCl. This phenomenon was also detected in the *csmA* disruptants. The authors also reported that the chitin content in the cell wall of the '*chsD*' disruptant was reduced to approximately 60% of that of the wild-type strain [14].

The Af*chsE* gene encoding class V chitin synthase was also cloned from *A. fumigatus* [18]. The disruptants of this gene showed abnormal hyphal structures and reduced conidiation efficiency. The disruptants formed abnormally swollen structures along hyphae and conidiophores. The virulence of this mutant was not significantly reduced when compared with the wild-type parental strain in a mouse model of pulmonary aspergillosis [18]. Thus, class V chitin synthases or enzymes with a chitin synthase domain of this class play some role in hyphal growth and conidiation.

Another type of chitin synthase-like gene (*chsD*), which does not belong to any class, was cloned from *A. fumigatus* and disrupted. The disruptant showed no morphological change, but the chitin content of the cell wall of the disruptant was reduced to 80% of the parental strain [17].

Double Disruption of *chs* Genes

As described above, the phenotype of strains carrying disruptions of *chsA, chsC*, or *chsD* was indistinguishable from that of the wild-type strain. Therefore, we constructed disruptants of two of three genes, *chsA, chsC* and *chsD*, in combinations. The double disruptants of *chsC* and *chsD* (CD disruptants) showed no phenotypic change, while the conidiation efficiency of the double disruptants of *chsA* and *chsD* (AD disruptants) was as low as 10% [12] and that of the double disruptants of *chsA* and *chsC* (AC disruptants) was less than 0.1% of that of the wild-type strain [Fujiwara et al., submitted].

Furthermore, AC disruptants showed pleiotropic defects. They were sensitive to high concentrations of salts, SDS, Calcofluor white, Congo red, and chitin synthase inhibitors, such as nikkomycin Z, polyoxin B and polyoxin D. Abnormal structures of conidiophores of AC disruptants were detected by scanning electron microscopy (fig. 6). AC disruptants formed the branching chains of elongated sterigmata, and only few conidia at the tips of sterigmata which were ambiguously separated. These sterigmata would be defined as metulae because they potentially branched. Calcofluor white staining of these structures showed that each compartment was septated (fig. 7) and contained one or two nuclei. Secondary conidiophores derived from a vesicle were found

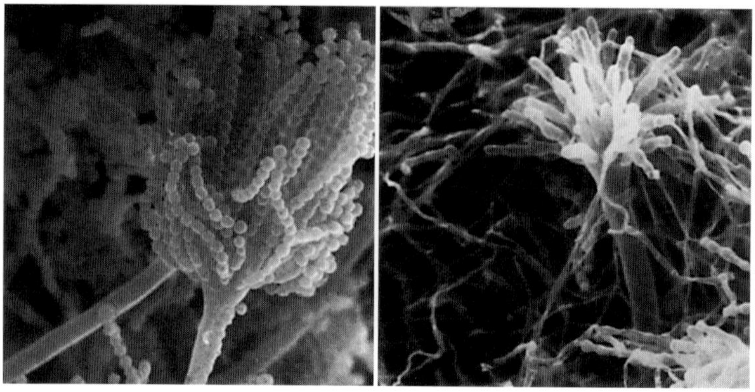

a b

Fig. 6. Structures of conidiophores of the wild-type strain, and the *chsA* and *chsC* double disruptant. Scanning microscopic analysis was done with the wild-type strain *a* and the *chsA* and *chsC* double disruptant *b*. Bar = 10 μm.

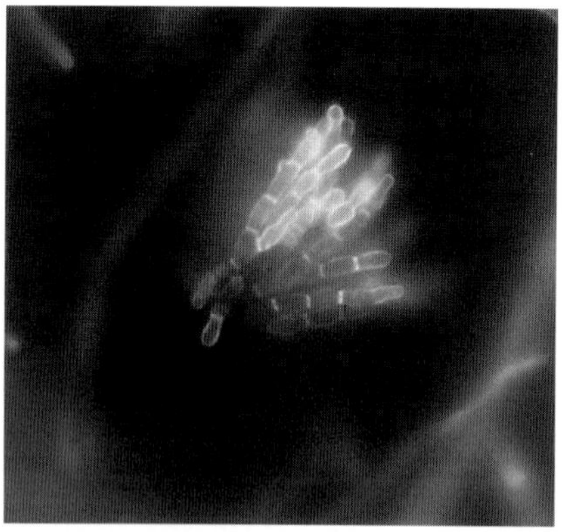

Fig. 7. Calcofluor white staining of conidiophores of the *chsA* and *chsC* double disruptant. Bar = 10 μm.

occasionally. In addition, no sexual reproductive cells, cleistothecia, were observed in the double disruptants. The abnormal structures of conidiophores and the inability of cleistothecia formation were also observed in the *medA* mutant of *A. nidulans* [25, 26]. These results suggest that *chsA* and *chsC* have important but at least partially redundant functions for tip growth of hyphae and formation of conidiophores.

Double disruptants of *csmA* and *chsD* were also constructed. Their phenotypes were almost the same as those observed for the *csmA* single disruptant. This finding suggested that there is obviously no functional redundancy of *csmA* and *chsD* [Fujiwara et al., in prep.].

The double disruptant of Af*chsC* and Af*chsG* showed almost the same phenotype as the Af*chsG* single disruptant and caused pulmonary disease in neutropenic mice at a reduced level [15].

Conclusions

ChsB and CsmA of *A. nidulans* have important functions in both formation of the normal morphology of hyphae and differentiation of conidiophores. ChsA and ChsC have important, but redundant functions in hyphal cell wall integrity and differentiation. These results suggest that spatially and strictly regulated activities of these chitin synthases are essential for normal morphogenesis of hyphae and formation of conidiophores of filamentous fungi. Therefore, detailed analyses of the function of these chitin synthases in the future will contribute to elucidate the mechanism of morphogenesis of filamentous fungi and are required to develop antifungal drugs with chitin synthases as their targets.

References

1 Bulawa CE: Genetics and molecular biology of chitin synthesis in fungi. Annu Rev Microbiol 1993; 47:505–534.

2 Cabib E, Silverman SJ, Shaw JA: Chitinase and chitin synthase 1: Counterbalancing activities in cell separation of *Saccharomyces cerevisiae*. J Gen Microbiol 1992;138:97–102.

3 Nagahashi S, Sudoh M, Ono N, Sawada R, Yamaguchi E, Uchida Y, Mio T, Takagi M, Arisawa M, Yamada-Okabe H: Characterization of chitin synthase 2 of *Saccharomyces cerevisiae*. J Biol Chem 1995;270:13961–13967.

4 Ford RA, Shaw JA, Cabib E: Yeast chitin synthase 1 and 2 consist of a non-homologous and dispensable N-terminal region and of a homologous moiety essential for function. Mol Gen Genet 1996;252:420–428.

5 Uchida Y, Shimmi O, Sudoh M, Arisawa M, Yamada-Okabe H: Characterization of chitin synthase 2 of *Saccharomyces cerevisiae*. II. Both full size and processed enzymes are active for chitin synthesis. J Biochem 1996;119:659–666.

6 DeMarini DJ, Adams AEM, Fares H, Virgilio CD, Valle G, Chuang JS, Pringle JR: A septin-based hierarchy of proteins required for localized deposition of chitin in the *Saccharomyces cerevisiae*. J Cell Biol 1997;139:75–93.

7 Santos B, Snyder M: Targeting of chitin synthase 3 to polarized growth sites in yeast requires Chs5p and Myo2p. J Cell Biol 1997;136:95–110.

8 Chuang JS, Schekman RW: Differential trafficking and timed localization of two chitin synthase proteins, Chs2p and Chs3p. J Cell Biol 1996;135:597–610.

9 Ziman M, Chuang JS, Schekman RW: Chs1p and Chs3p, two proteins involved in chitin synthesis, populate a compartment of the *Saccharomyces cerevisiae* endocytic pathway. Mol Biol Cell 1996; 7:1909–1919.

10 Yanai K, Kojima N, Takaya N, Horiuchi H, Ohta A, Takagi M: Isolation and characterization of two chitin synthase genes from *Aspergillus nidulans*. Biosci Biotech Biochem 1994;58:1828–1835.

11 Motoyama T, Kojima N, Horiuchi H, Ohta A, Takagi M: Isolation and characterization of a chitin synthase gene (*chsC*) of *Aspergillus nidulans*. Biosci Biotech Biochem 1994;58:2254–2257.

12 Motoyama T, Fujiwara M, Kojima N, Horiuchi H, Ohta A, Takagi M: The *Aspergillus nidulans chsD* gene encodes a chitin synthase which has a redundant function with the *chsA* gene product in conidiophore formation. Mol Gen Genet 1996;251:442–450, 1997;253:520–528.

13 Fujiwara M, Horiuchi H, Ohta A, Takagi M: A novel fungal gene encoding chitin synthase with a myosin motor-like domain. Biochem Biophys Res Commun 1997;236:75–78.

14 Specht CA, Liu Y, Robbins PW, Bulawa CE, Iartchouk N, Winter KR, Riggle PJ, Rhodes JC, Dodge CL, Culp DW, Borgia P: The *chsD* and *chsE* genes of *Aspergillus nidulans* and their roles in chitin synthesis. Fungal Genet Biol 1996;20:153–167.

15 Mellado E, Aufauvre-Brown A, Specht CA, Robbins PW, Holden DW: A multigene family related to chitin synthase genes of yeast in the opportunistic pathogen *Aspergillus fumigatus*. Mol Gen Genet 1995;246:353–359.

16 Mellado E, Aufauvre-Brown A, Gow NA, Holden DW: The *Aspergillus fumigatus chsC* and *chsG* genes encode class III chitin synthases with different functions. Mol Microbiol 1996;20:667–679.

17 Mellado E, Specht CA, Robbins PW, Hoden DW: Cloning and characterization of *chsD*, a chitin synthase-like gene of *Aspergillus fumigatus*. FEMS Microbiol Lett 1996;143:69–76.

18 Aufauvre-Brown A, Mellado E, Gow NAR, Holden DW: *Aspergillus fumigatus chsE:* A gene related to *CHS3* of *Saccharomyces cerevisiae* and important for hyphal growth and conidiophore development but not pathogenicity. Fungal Genet Biol 1997;21:141–152.

19 Tatsuno K, Yamada-Okabe H, Takagi M, Arisawa M, Sudoh M: Properties of yeast expressed *Aspergillus nidulans* chitin synthase B which is essential for hyphal growth. FEMS Microbiol Lett 1997;149:279–284.

20 Beth Din A, Yarden O: The *Neurospora crassa chs-2* gene encodes a non-essential chitin synthase. Microbiology 1994;140:2189–2197.

21 Beth Din A, Specht CA, Robbins PW, Yarden O: *chs-4*, a class IV chitin synthase gene from *Neurospora crassa*. Mol Gen Genet 1996;250:214–222.

22 Yarden O, Yanofsky C: Chitin synthase 1 plays a major role in cell wall biogenesis in *Neurospora crassa*. Genes Dev 1991;5:2420–2430.

23 Waring RB, May GS, Morris NR: Characterization of an inducible expression system in *Aspergillus nidulans* using *alcA* and tubulin-coding genes. Gene 1989;79:119–130.

24 Borgia PT, Iartchouk N, Riggle PJ, Winter KR, Koltin Y, Bulawa CE: The *chsB* gene of *Aspergillus nidulans* is necessary for normal hyphal growth and development. Fungal Genet Biol 1996;20: 193–203.

25 Aguirre J: Spatial and temporal controls of the *Aspergillus brlA* developmental regulatory gene. Mol Microbiol 1993;8:211–218.

26 Busby TM, Miller KY, Miller BL: Suppression and enhancement of the *Aspergillus nidulans medusa* mutation by altered dosage of the *bristle* and *stunted* genes. Genetics 1996;143:155–163.

Masamichi Takagi, Department of Biotechnology, The University of Tokyo,
Bunkyo-ku, Tokyo 113 (Japan)
Fax +81 3 3812 9246

Brakhage AA, Jahn B, Schmidt A (eds): *Aspergillus fumigatus*.
Contrib Microbiol. Basel, Karger, 1999, vol 2, pp 205–215

Pigment Biosynthesis and Virulence

Axel A. Brakhage[a], *Kim Langfelder*[a], *Gerhard Wanner*[b], *Axel Schmidt*[c], *Bernhard Jahn*[d]

[a] Institut für Mikrobiologie und Genetik, Technische Universität Darmstadt;
[b] Botanisches Institut, Universität München;
[c] Institut für Chemotherapie, Wuppertal, und
[d] Institut für Medizinische Mikrobiologie, Johannes Gutenberg-Universität, Mainz, Deutschland

The importance of *Aspergillus fumigatus* as an opportunistic human-pathogenic fungus has been outlined in several chapters of this volume [1, 2]. One of the important questions concerning *A. fumigatus* is the identification of virulence factors. Normally, *A. fumigatus* is a saprobic fungus, associated with decaying organic matter such as compost and hay. It plays an essential role in recycling carbon and nitrogen sources [3, 4]. Thus it is doubtful whether *A. fumigatus* has specific virulence factors [5, 6]. It is more likely that the fungus has certain physiological characteristics that enable it to avoid or suppress the residual immune system in immunocompromised patients, making it an aggressive opportunistic pathogen [5].

Recently, such a physiological factor, which is associated with the conidia, was identified. The grey-green conidia appear to contain 1,8-dihydroxynaphthalene (DHN)-melanin which protects them, at least in part, against the attack of immune effector cells [7–10]. In the remainder of this article, the arguments and the experimental evidence in favour of this conclusion are given.

Importance of the *A. fumigatus* Conidial Pigment for Virulence and Its Possible Mechanisms in Protecting Conidia against the Attack of Immune Effector Cells

The survival of conidia in the host and the onset of germination are prerequisite for developing disease. This indicates that some factors attributed to the conidia of *A. fumigatus* help the fungus to survive the attack of the residual host immune system. Because of their abundance in air, *A. fumigatus* conidia are continuously inhaled during routine daily activities. Mucociliary clearance and phagocytic defense normally prevent the disease, but a severely depressed immune system provides an opportunity for conidia to germinate and invade lung tissue [1, 2]. We reasoned that one of the factors that contributed to conidial survival might be the pigment of the grey-green spores [7]. This reasoning was also based on results obtained from other fungi. One example is *Cryptococcus neoformans*, a fungus with a pathogenic yeast phase, which causes life-threatening infections particularly in AIDS patients [11]. It has been shown that production of a melanin-like pigment is a virulence-determining factor in *C. neoformans* [12–15]. One mechanism by which pigment might contribute to virulence derives from its ability to confer some resistance to reactive oxygen species (ROS), a major host antimicrobial effector system also active against *Aspergillus* conidia [16–18]. For the dematiacous fungus *Wangiella dermatitidis*, a main causative agent of phaeohyphomycosis, a similar biological function of a fungal pigment was demonstrated [19, 20].

In order to investigate the importance of the conidial pigmentation for the virulence of *A. fumigatus*, a mutant of *A. fumigatus* was isolated by UV mutagenesis which lacked the ability to produce the conidial pigment and therefore had white (W) conidia [7]. In an intravenous murine infection model, the virulence of conidia of the W strain was significantly reduced to approximately 20–50% of that observed with wild-type (WT) conidia [7]. This suggested that the conidial pigment was indeed important for virulence.

To elucidate the mechanisms by which the pigment might protect conidia against the attack of immune effector cells, a series of in vitro experiments was performed. When conidia were incubated with reactive oxygen species (ROS) like NaOCl or hydrogen peroxide, it became evident that W conidia were about 10-fold more sensitive to ROS than WT conidia. This finding indicated that the pigment gives *A. fumigatus* conidia some protection against ROS. In addition, W conidia were more efficiently damaged by human monocytes than WT conidia in vitro, suggesting a possible link between conidial pigment and the survival of conidia in monocytes [7].

The conidia were also tested for their ability to cause ROS production in human monocytes and polymorphonuclear leukocytes (PMN). The release of ROS was assessed by luminol-mediated chemiluminescence. Conidia of the WT strain only led to a small release of ROS, which agreed well with previous results reported by Levitz and Diamond [21]. By contrast, conidia of the W mutant caused a 10-fold higher ROS release by both cell types [7]. Before discussing these results, it must be mentioned that the analysis of the surface of conidia by scanning electron microscopy led to an unexpected finding. Whereas WT conidia showed a distinct ornamentation on their surface, W conidia had a smooth surface (fig. 1). Hence, besides a lack of pigmentation, the mutants showed another phenotype which might be important for the interplay between conidia and immune effector cells. It was also conceivable that there was no direct link between the lack of pigmentation and ornamentation of the conidium surface. Both phenotypes could be due to different mutations present in the W mutant. However, a classical reversion analysis indicated that all phenotypes found in the W mutant were due to a single mutation [7]. This was confirmed by the cloning of a gene complementing both phenotypes of the W mutant [9, 10]. These findings provided evidence that the ornamentation of WT conidia was dependent on the presence of the pigment.

There are two possible explanations for the observation that the W conidia induced a 10-fold greater release of ROS when coincubated with PMN and monocytes than the WT conidia. The first explanation is that conidia of both strains induced an identical oxidative burst. However, the resulting ROS were scavenged by the pigment present in WT conidia and thus were not detected by chemiluminescence. For some fungal pigments, their abilities to scavenge ROS have been already shown [14, 19] and here, this assumption was supported by the observation that W conidia were about 10-fold more sensitive to both hydrogen peroxide and NaOCl than WT conidia. This indicated that ROS scavenging also occurs in pigmented *A. fumigatus* conidia and thus confers a degree of protection against damage caused by ROS. The second explanation is that W conidia were better recognised by phagocytes because the conidial surface of the W mutant markedly differed from that of WT conidia. This might result in a more efficient and stronger activation of phagocytes which would explain the observed increase in production of ROS when the immune cells were challenged with W mutant conidia.

The latter assumption was also supported by the findings that conidia of a W mutant independently isolated by Tsai et al. [8] exhibited significantly elevated complement component C3 binding capacity compared with wild-type conidia. Inefficient deposition of complement component C3 on wild-type *A. fumigatus* conidia has previously been demonstrated [22]. Also, there is a good correlation between binding of C3 molecules per unit conidial surface and pathogenicity, i.e. highly pathogenic *A. fumigatus* and *A. flavus* bound fewer C3 molecules than the nonpathogenic species *Emericella* (*Aspergillus*) *nidulans* [23]. Taken together, these results strongly suggest that conidial pigmentation and/or associated structures contribute to the virulence of *A. fumigatus.* As both mechanisms, i.e. lack of quenching ability as well as activation of the host immune defense mechanisms by altered conidial surface, are functionally independent from each other, the observed difference in ROS release may well result from a combined effect of both mutant properties.

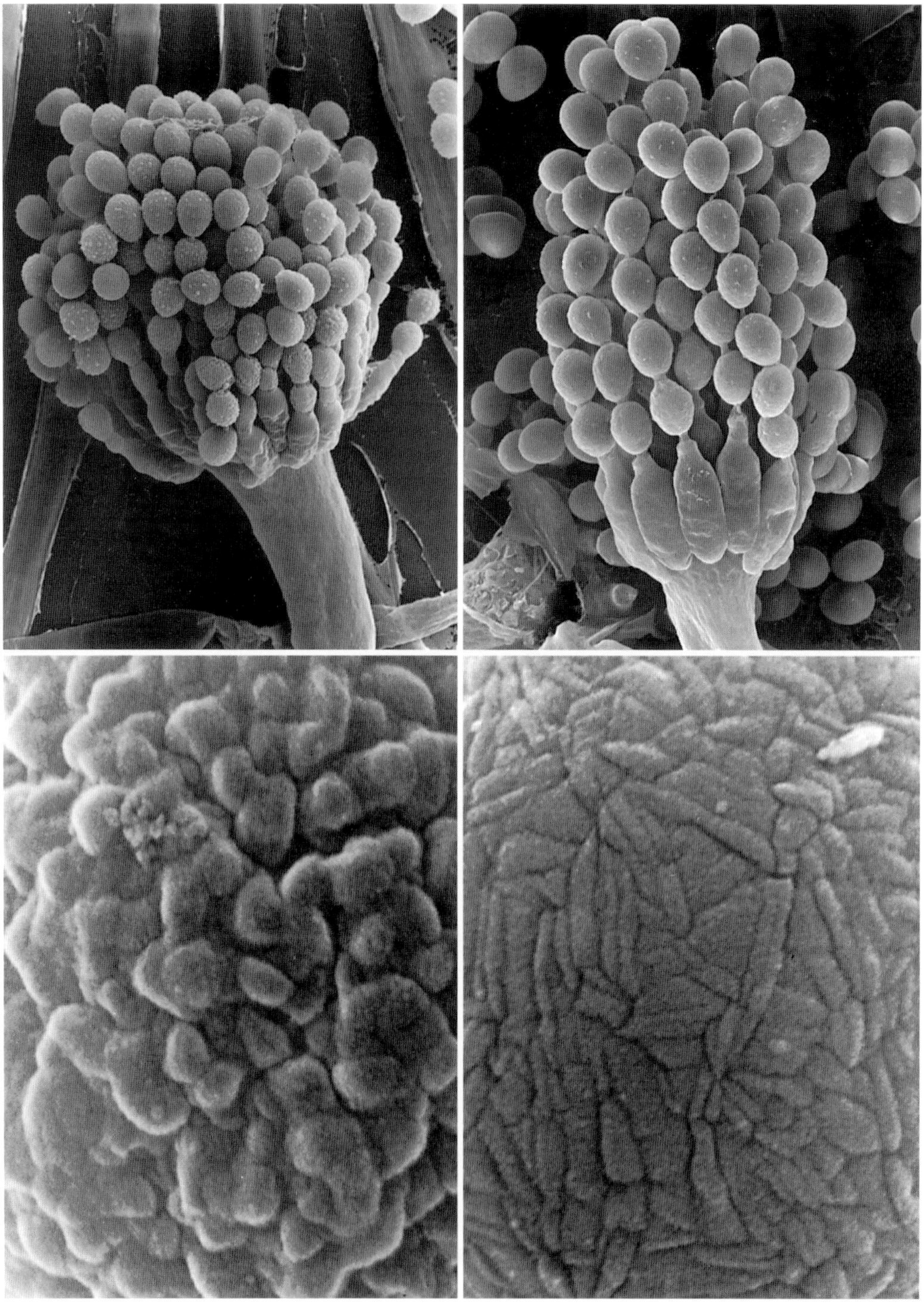

A B a b

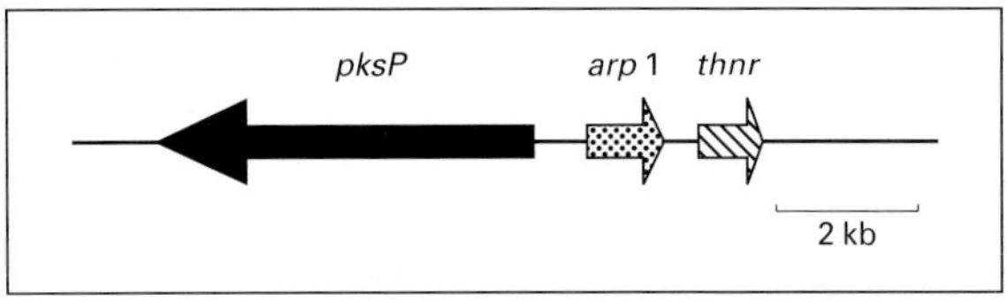

Fig. 2. Gene cluster containing *pksP* (*alb1*), *arp1*, and a putative 1,3,6,8-tetrahydroxy-napthalene (THN) reductase gene designated *thnr*. *pksP* (*alb1*) encodes a putative polyketide synthase [9, 10], *arp1* scytalone dehydratase [8] and *thnr* 1,3,6,8-tetrahydroxy-napthalene reductase [Langfelder and Brakhage, unpubl.]. The possible functions of the gene products in the biosynthesis of DHN-melanin are shown in figure 3. The transcriptional direction of genes is marked by arrows.

A Polyketide Synthase Gene Complemented the W Mutant Strain to Wild Type

To get further insight into the role of the conidial pigment in the virulence of *A. fumigatus*, a gene complementing the W mutant to the wild-type phenotype was independently isolated by Langfelder et al. [9] and Tsai et al. [10]. Although there were marginal differences in the DNA sequences reported, both research groups seem to have isolated the same gene. Since its deduced amino acid sequence exhibits a high degree of similarity to polyketide synthases, the gene was designated *pksP* for polyketide synthase involved in pigmentation (EMBL accession No. Y17317) [9], and *alb1* (Genbank accession No. g3136091) [10]. In the remainder of the article the name *pksP* is used. The *pksP* gene has a size of 6,660 bp. The deduced ORF of the *A. fumigatus* PksP protein is encoded by 6,660 bp interrupted by four introns of 47, 47, 55 and 73 bp length (fig. 2). The location of introns was deduced by identification of consensus splicing sites according to Gurr et al. [24] and comparison to the *E. nidulans wA* gene. The deduced PksP protein is composed of 2,146 amino acids with a calculated molecular mass of 234.3 kD [9].

W mutant strains complemented to the grey-green phenotype of WT conidia using the *pksP* gene also showed a wild-type ornamentation of the conidial surface. In addition, they triggered the same ROS release by PMN as wild-type conidia which was 10-fold less than the amount of ROS released by PMN when challenged with W conidia. In an intravenous murine infection model, the virulence of transformants complemented to the wild type was restored and corre-

Fig. 1. Scanning electron microscopic photographs of *A. fumigatus* wild type (WT) (*A*) and W mutant (*B*). Pictures marked by lower case letters indicate enlargements of the conidial surface of the WT (*a*) and W mutant (*b*).

sponded to that of WT conidia [9]. In addition, the disruption of the *pksP* (*alb1*) gene resulted in the phenotypes of conidia characteristic of W conidia, i.e. white conidia, lack of ornamentation and reduced virulence of conidia of the disruptant strain in an intravenous murine infection model [10].

As mentioned above, the deduced amino acid sequence of PksP showed extensive sequence similarities with fungal polyketide synthases, in particular with the WA protein of *E. nidulans* (60.8% identical amino acids) and Pksl of *Colletotrichum lagenarium*, indicating that *pksP* encodes a polyketide synthase. According to the definition of polyketide synthases [25], the *pksP* gene encodes a typical fungal type I polyketide synthase. PksP shares β-ketotransferase, acyltransferase, acyl carrier protein and thioesterase motifs with other fungal polyketide synthases. It is therefore likely that *pksP* encodes all or part of a polyketide synthase involved in the formation of the *A. fumigatus* conidial pigment [9, 10].

PksP Is Part of a Cluster of 1,8-Dihydroxynaphthalene (DHN)-Melanin Biosynthesis Genes

The *pksP* gene is part of a cluster [9, 10] (fig. 2). Upstream of *pksP* the *arp1* gene is located. *Arp1* has a size of 670 bp. The ORF of *arp1* is interrupted by two introns and encodes a protein of 168 amino acids. The *arp1* gene product is very similar to scytalone dehydratase, an enzyme involved in 1,8-dihydroxynaphthalene (DHN)-melanin synthesis in the plant-pathogenic fungi *C. lagenarium* and *Magnaporthe grisea* [8] (fig. 3). Disruption of *arp1* resulted in reddish-pink conidia, indicating its involvement in the synthesis of the conidial pigment, and increased C3 binding. Both *arp1* and *pksP* have been shown to be developmentally regulated [8, 10].

Downstream of *arp1* is a gene located with high similarity to 1,3,6,8-tetrahydroxy-napthalene (THN) reductase genes. It was therefore designated *thnr* [Langfelder and Brakhage, unpubl.] (fig. 2). Based on previous results on the DHN-melanin [26], it is very likely that the *thnr* gene product is also involved in the DHN-melanin pathway (fig. 3).

A. fumigatus Conidia Contain DHN-Melanin in Contrast to Conidia of the Non-Pathogenic *E. nidulans*

There are very few publications reporting the isolation of *E. nidulans* from material isolated from patients. Since there is no evidence of pathogenicity of *E. nidulans* per se, we regard this fungus as nonpathogenic. Mutants of the

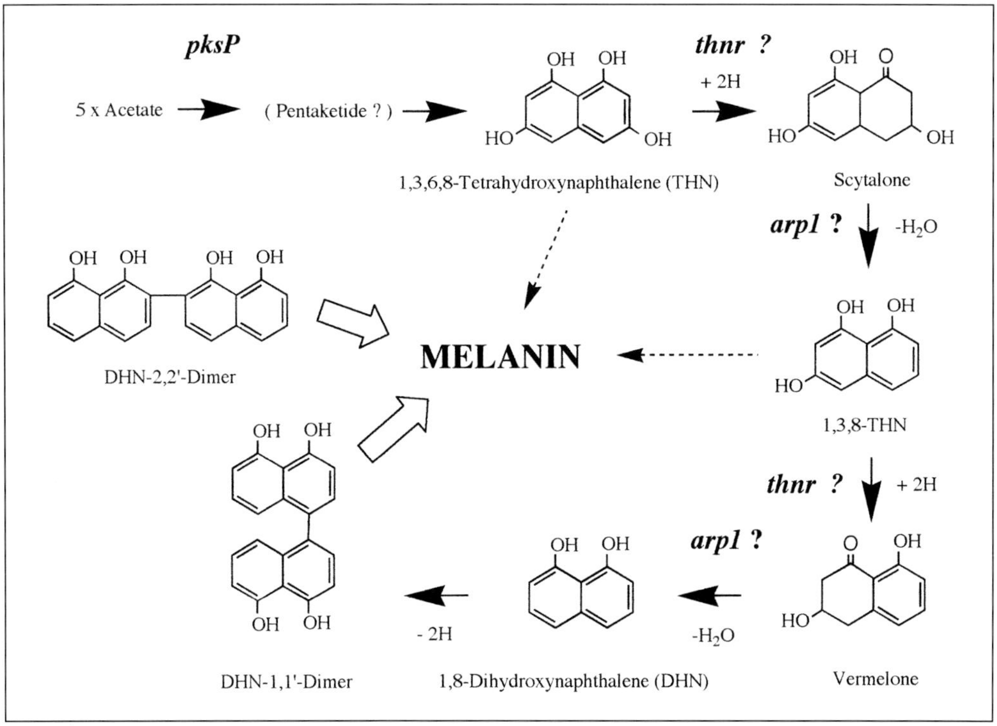

Fig. 3. Current view of the DHN-melanin biosynthesis pathway according to Bell and Wheeler [26]. The reactions which are assumed to be catalysed by the products of the *A. fumigatus pksP, arp1* and *thnr* gene are indicated.

nonpathogenic *E. nidulans* were previously isolated which also produce white conidia. One of these mutants (WA) was complemented by the *wA* gene of *E. nidulans* [27, 28]. Since pigment synthesis in *E. nidulans* involves a polyketide pathway [29], WA seems to contribute to conidial pigment biosynthesis [30]. It might have a similar function to the PksP of *A. fumigatus*, although major differences between the pigment biosyntheses of the two fungal species appear to exist (see below). PksP also showed a high degree of similarity to Pks1 of the plant-pathogenic fungus *C. lagenarium. pks1* encodes a polyketide synthase which is required for the biosynthesis of DHN-melanin [31] which is essential for the function of the appressorium in penetration of host plants [32]. It is thus regarded as a virulence factor. In agreement with this conclusion, albino strains lacking the *pks1* gene have reduced penetration ability and thus attenuated pathogenicity on the host plant [33]. Genes encoding enzymes of the DHN-melanin pathway have been cloned from several plant pathogenic fungi,

Table 1. Genes identified until now of DHN-melanin biosynthesis of some plant-pathogenic fungi and *A. fumigatus*

Gene	Assumed product	Ref. No.
Aspergillus fumigatus		
pksP (*albl*)	polyketide synthase	7, 9, 10
arpl	scytalone dehydratase	8
thnr	1,3,6,8-THN reductase	Langfelder and Brakkhage, in prep.
Colletotrichum lagenarium		
PKS1	polyketide synthase	31
SCD1	scytalone dehydratase	36
THR1	1,3,6,8-THN reductase	37
Magnaporthe grisea		
T_4HN	1,3,6,8-THN reductase	38

but for none of the fungi have all the genes involved in the biosynthesis pathway been cloned (table 1).

The product of the *C. lagenarium* Pksl is a pentaketide [26]; that of *E. nidulans* WA or *A. fumigatus* PksP has not yet been identified. It is difficult to predict the structures of the products of the latter two enzymes, because the involvement of other polyketide synthases in the generation of the final pigment cannot yet be excluded.

The DHN-melanin pathway has been examined in *A. fumigatus* [34]. The fungicide tricyclazole specifically inhibits the reductases involved in DHN-melanin biosynthesis in black fungi [29]. The inhibition of conidial pigmentation of *A. fumigatus* by tricyclazole suggested that the grey-green pigment of *A. fumigatus* was synthesized via a pathway similar to the DHN-melanin pathway of *C. lagenarium* and *M. grisea* [8, 29]. This was also supported by the cloning of the *arp1* gene of *A. fumigatus* which is involved in conidial pigment synthesis and putatively encodes a scytalone dehydratase [8] (fig. 3). Similar enzymes have been found in *C. lagenarium* and *M. grisea* and shown to be involved in DHN-melanin synthesis [29]. Interestingly, in contrast to *A. fumigatus*, the *E. nidulans* conidial pigmentation was not inhibited by tricyclazole [29] and, in addition, Tsai et al. [8] could not detect any homologues of *arp1* in *E. nidulans* with low stringency hybridisation, suggesting dissimilar pathways in *A. fumigatus* and *E. nidulans* pigment synthesis. The identification of the *thnr* gene encoding a putative THN reductase gene whose product is most likely also involved in the DHN-melanin pathway (fig. 3) further supports the assumption that the pigment of *A. fumigatus* is synthesized via the DHN-melanin pathway.

For several human pathogenic fungi, including *Cryptococcus neoformans* and *Wangiella dermatitidis*, melanin was shown to be important for virulence [12–14, 19, 29, 35]. Since the DHN-melanin-containing conidial pigment of *A. fumigatus* is also important for virulence and differs from the conidial pigment of the nonpathogenic *E. nidulans*, it seems to represent a characteristic feature of *A. fumigatus* which is a prerequisite for the virulence of this fungus. Hence, it will be of considerable interest to analyze the differences in the types of melanin and the regulation of their biosyntheses between the pathogenic *A. fumigatus* and nonpathogenic *E. nidulans*. This can be expected to have a major impact on the understanding of the molecular basis of the virulence of *A. fumigatus*.

Conclusion

Conidia of *A. fumigatus* contain a grey-green pigment which is important for the virulence of the fungus. Possible mechanisms by which the pigment protects conidia against killing by the host defense include the scavenging of ROS produced as antimicrobial defense by immune effector cells and/or the reduced recognition of pigmented conidia by the host immune system, as demonstrated by reduced binding of complement C3 to WT conidia compared with W or Arpl mutants.

The reduced recognition of WT conidia compared with W conidia by the immune system might also be connected to the pigment since WT conidia display a characteristic surface structure which is lacking in the pigmentless conidia of the W mutant. In addition, genetic and biochemical data strongly suggest that the conidial pigment of *A. fumigatus* contains DHN-melanin which is absent in the nonpathogenic *E. nidulans* and therefore seems to represent a physiological factor of major importance for the virulence of *A. fumigatus* and might help to explain why this fungus has the potential to be pathogenic.

Acknowledgements

The work in the author's laboratories was supported by the Deutsche Forschungsgemeinschaft (Br 1130/5-1, to A. A. B.) and in part by SFB369 (to G.W.).

References

1 Rüchel R, Reichard U: Pathogenesis and clinical presentation of aspergillosis; in Brakhage AA, Jahn B, Schmidt A (eds): *Aspergillus fumigatus:* Biology, Clinical Aspects and Molecular Approaches to Pathogenicity. Contrib Microbiol. Basel, Karger, 1999.

2 Ellis M: Therapy of *Aspergillus fumigatus*-related diseases; in Brakhage AA, Jahn B, Schmidt A (eds): *Aspergillus fumigatus:* Biology, Clinical Aspects and Molecular Approaches to Pathogenicity. Contrib Microbiol. Basel, Karger, 1999.

3 Mullins J, Harvey R, Seaton A: Sources and incidence of airborne *Aspergillus fumigatus* (Fres.). Clin Allergy 1976;6:209–217.

4 Debeaupuis J-P, Sarfati J, Chazalet V, Latgé J-P: Genetic diversity among clinical and environmental isolates of *Aspergillus fumigatus*. Infect Immun 1997;65:3080–3085.

5 Holden DW, Tang CM, Smith JM: Molecular genetics of *Aspergillus* pathogenicity. Antonie Van Leeuwenhoek 1994;65:251–255.

6 Monod M, Jaton-Ogay K, Reichard U: *Aspergillus fumigatus*-secreted proteases as antigenic molecules and virulence factors; in Brakhage AA, Jahn B, Schmidt A (eds): *Aspergillus fumigatus:* Biology, Clinical Aspects and Molecular Approaches to Pathogenicity. Contrib Microbiol. Basel, Karger, 1999.

7 Jahn B, Koch A, Schmidt A, Wanner G, Gehringer H, Bhakdi S, Brakhage AA: Isolation and characterisation of an *Aspergillus fumigatus* mutant strain with pigmentless conidia and reduced virulence. Infect Immun 1997;65:5110–5117.

8 Tsai H-F, Washburn RG, Chang YC, Kwon-Chung KJ: *Aspergillus fumigatus arp1* modulates conidial pigmentation and complement deposition. Mol Microbiol 1997;26:175–183.

9 Langfelder K, Jahn B, Gehringer H, Schmidt A, Wanner G, Brakhage AA: Identification of a polyketide synthase gene (*pksP*) of *Aspergillus fumigatus* involved in conidial pigment biosynthesis and virulence. Med Microbiol Immunol 1998;187:79–89.

10 Tsai H-F, Yun CC, Washburn RG, Wheeler MH, Kwong-Chung KJ: The developmentally regulated *alb1* gene of *Aspergillus fumigatus:* Its role in modulation of conidial morphology and virulence. J Bacteriol 1998;180;3031–3038.

11 Zuger A, Louie E, Holzman RS, Simberkoff MS, Rahal JJ: Cryptococcal disease in patients with the acquired immunodeficiency syndrome: Diagnostic features and outcome of treatment. Ann Intern Med 1986;104:234–240.

12 Kwon-Chung KJ, Polacheck I, Popkin TJ: Melanin-lacking mutants of *Cryptococcus neoformans* and their virulence for mice. J Bacteriol 1982;150:1414–1421.

13 Kwon-Chung KJ, Rhodes JC: Encapsulation and melanin formation as indicators of virulence in *Cryptococcus neoformans*. Infect Immun 1986;51:218–223.

14 Wang Y, Aisen P, Casadevall A: *Cryptococcus neoformans* melanin and virulence: Mechanism of action. Infect Immun 1995;63:3131–3136.

15 Alspaugh JA, Perfect JR, Heitman J: *Cryptococcus neoformans* mating and virulence are regulated by the G-protein α subunit GPA1 and cAMP. Genes Dev 1997;11:3206–3217.

16 Babior BM: Oxygen-dependent microbial killing by phagocytes (first of two parts). N Engl J Med 1978;298:659–668.

17 Washburn RG, Gallin JI, Bennett JE: Oxidative killing of *Aspergillus fumigatus* proceeds by parallel myeloperoxidase-dependent and -independent pathways. Infect Immun 1987;55:2088–2092.

18 Wang Y, Casadevall A: Susceptibility of melanized and nonmelanized *Cryptococcus neoformans* to nitrogen- and oxygen-derived oxidants. Infect Immun 1994;62:3004–3007.

19 Jacobson ES, Hove E, Emery HS: Antioxidant function of melanin in black fungi. Infect Immun 1995;63:4944–4945.

20 Jacobson ES, Tinnell SB: Antioxidant function of fungal melanin. J Bacteriol 1993;175:7102–7104.

21 Levitz SM, Diamond RD: Mechanisms of resistance of *Aspergillus fumigatus* conidia to killing by neutrophils in vitro. J Infect Dis 1985;152:33–42.

22 Kozel TR, Wilson MA, Farrell TP, Levitz SM: Activation of C3 and binding to *Aspergillus fumigatus* conidia and hyphae. Infect Immun 1989;57:3412–3417.

23 Henwick S, Hetherington SV, Patrick CC: Complement binding to *Aspergillus* conidia correlates with pathogenicity. J Lab Clin Med 1993;122:27–35.
24 Gurr SJ, Unkles SE, Kinghorn JR: The structure and organisation of nuclear genes of filamentous fungi; in Kinghorn JR (ed): Gene Structure in Eukaryotic Microbes. Oxford, IRL Press, 1987, pp 93–139.
25 Hutchinson CR, Fujii I: Polyketide synthase gene manipulation: A structure-function approach in engineering novel antibiotics. Annu Rev Microbiol 1995;49:201–238.
26 Bell AA, Wheeler MH: Biosynthesis and functions of fungal melanins. Annu Rev Phytopathol 1986; 24:411–451.
27 Mayorga ME, Timberlake WE: Isolation and molecular characterisation of the *Aspergillus nidulans wA* gene. Genetics 1990;126:73–79.
28 Tilburn J, Roussel F, Scazzocchio C: Insertional inactivation and cloning of the *wA* gene of *Aspergillus nidulans*. Genetics 1990;126:81–90.
29 Wheeler MH, Bell AA: Melanins and their importance in pathogenic fungi; in McGinnis MR (ed): Current Topics in Medical Mycology. New York, Springer, 1988, pp 338–387.
30 Mayorga ME, Timberlake WE: The developmentally regulated *Aspergillus nidulans wA* gene encodes a polypeptide homologous to polyketide and fatty acid synthases. Mol Gen Genet 1992;235:205–212.
31 Takano Y, Kubo Y, Shimizu K, Mise K, Okuno T, Furusawa I: Structural analysis of *PKS1*, a polyketide synthase gene involved in melanin biosynthesis in *Colletotrichum lagenarium*. Mol Gen Genet 1995;249:162–167.
32 Yamaguchi I, Kubo Y: Target sites of melanin biosynthesis inhibitors; in Koller W (ed): Target Sites of Fungicide Action. London, CRC Press, 1992, pp 101–118.
33 Kubo Y, Furusawa I: Melanin biosynthesis: Prerequisite for successful invasion of the plant host by appressoria of *Colletotrichum* and *Pyricularia*; in Cole GT, Hoch HC (eds): The Fungal Spore and Disease Initiation in Plants and Animals. New York, Plenum Publishing, 1991, pp 205–217.
34 Wheeler MH, Klich MA: The effects of tricyclazole, pyroquilon, phthalide, and related fungicides on the production of conidial wall pigments by *Penicillium* and *Aspergillus* species. Pestic Biochem Physiol 1995;52:125–136.
35 Dixon DM, Polak A, Szaniszlo PJ: Pathogenicity and virulence of wild-type and melanin-deficient *Wangiella dermatitidis*. J Med Vet Mycol 1987;25:97–106.
36 Kubo Y, Takano Y, Endo N, Yasuda N, Tajima S, Furusawa I: Cloning and structural analysis of the melanin biosynthesis gene *SCD1* encoding scytalone dehydratase in *Colletotrichum lagenarium*. Appl Environm Microbiol 1996;62:4340–4344.
37 Perpetua NS, Kubo Y, Yasuda N, Takano Y, Furusawa I: Cloning and characterisation of a melanin biosynthetic THR1 reductase gene essential for appressorial penetration of *Colletotrichum lagenarium*. Mol Plant-Microbe Interactions 1996;9:323–329.
38 Vidal-Cros A, Viviani F, Labesse G, Boccara M, Gaudry M: Polyhydroxynaphthalene reductase involved in melanin biosynthesis in *Magnaporthe grisea:* Purification, cDNA cloning and sequencing. Eur J Biochem 1994;219:985–992.

Prof. Dr. Axel A. Brakhage, Institut für Mikrobiologie und Genetik,
Technische Universität Darmstadt, Schnittspahnstrasse 10, D–64287 Darmstadt (Germany)
Tel. +49 6151 165566, Fax +49 6151 162956, E-Mail Brakhage@bio.tu-darmstadt.de

Author Index

Subject Index